Building Your Own Home

Building Your Own Home

A
Step-by-Step
Guide

WASFI YOUSSEF, Ph.D.

WILEY

John Wiley & Sons, Inc.
New York • Chichester • Brisbane • Toronto • Singapore

Publisher: Stephen Kippur
Editor: Katherine Schowalter
Managing Editor: Ruth Greif
Editing, Design, and Production: Hudson River Studio

Youssef, Wasfi.
    Building your own home : a step-by-step guide / Wasfi Youssef.
        p.   cm.
    Bibliography: p.
    ISBN 0-471-63562-6.   ISBN 0-471-63561-8 (pbk.)
    1. House construction. 2. House construction—Contracts and specifications. I. Title.
    TH4811.Y68 1988
    690'.837—dc19   88-14910
                                                                        CIP

20   19   18   17   16   15

TO
THERESA

Preface

Building one's own home is a dream that many people would like to see come true. However, to be successful in this endeavor, one must have a strong background not only in architecture and construction but also in the legal, financial, and administrative aspects. This book explains logically and in detail the steps involved in building a house from scratch, starting with buying a lot and ending with obtaining the certificate of occupancy. It is written simply so that it should be clear to both laypeople and professionals.

Building Your Own Home: A Step-by-Step Guide is especially beneficial to those who intend to build a new home themselves or to make additions, renovations, or improvements to their existing homes. The legal and financial chapters are of great importance to individuals who are contracting with a developer or a general contractor to build them a turn-key (complete) house, as well as to those who intend to buy a previously owned home. Knowing about different types of deeds and deed restrictions, types of ownership, protecting the title, the contract of sale of real estate, types of mortgages, mortgage applications, and deed recording *before* paying a deposit can prevent costly mistakes. The reader will become familiar with the contents of the various documents he or she will be required to sign before and at the closing.

Because the book is diversified, it can be a useful reference to architects, engineers, developers, real estate professionals, and home inspectors and appraisers.

The ideas, recommendations, and conclusions presented in this text are subject to local, state, and federal laws and regulations and their revisions; and to the local building codes and zoning ordinances; and to the nature and topography of your particular lot. This book should not be used as a substitute for competent legal, financial, and technical advice.

Wasfi Youssef

Acknowledgments

I am grateful for the interest and encouragement I have received from Katherine R. Schowalter, editor at John Wiley & Sons, Inc.

Both thanks and gratitude are tendered to my wife Theresa for her unequivocal support and unfailing help during the writing of this book.

Contents

CHAPTER 1

Introduction

*T*his book explains step by step what you need to know in order to build a home. Here is a description of the way it is organized.

Chapters 2 through 18 are devoted to the legal, financial, and administrative steps that precede construction. The factors that should be considered when buying a lot are described in detail. Special attention is given to subsurface water problems that may not be visible at the time of buying the lot but may crop up in the basement after you've built and moved into the house.

Different types of contracts are described. Because of its significance, the essentials of the contract of sale of real estate are presented in detail. The difference between the offer to purchase and a contract of sale are explained. A chapter is devoted to property description and land survey, which define precisely the size and location of the property you are buying.

The subject of liens is fully explained in another chapter. Few people have ever heard of the mechanic's lien and fewer know how it works. However, you should be aware of this type of lien at the outset since, if attached to

your house while it is under construction, it can keep you from obtaining subsequent loan installments. Furthermore, you cannot obtain a mortgage loan or sell the house if it has a mechanic's lien attached to it. Ways of protecting yourself against this lien are presented in detail.

The title, public recording, and recording statutes of different states are explained in another chapter. You will also learn how a title search is conducted and how to ensure that the property you are buying will not involve you in litigation.

A deed is a document by which the ownership to a property is transferred. There are different types of deeds: warranty deed with full covenants, bargain and sale deed, quitclaim deed, tax deed, referee's deed, sheriff's deed, and trustee deed. The legal rights and grantees contained in each of these deeds are explained.

Land use can be restricted by the zoning ordinances, deed restrictions, or both. Any landowner can impose restrictions on land use by subsequent owners by including them in the deed when the property is sold.

Mortgage law, mortgage documents,

types of mortgages, default, and fore-closure are explained in two chapters. The rights of the lender with regard to the mortgaged property in different states are fully explained.

Financing and lending practices, primary mortgage market (the lending institutions), secondary mortgage market, and unconventional (also called creative) financing are explained at length. This information should help you in shopping for a mortgage loan. One article is devoted to house appraisal; the particulars that affect the price of a property are listed. These particulars can be used as a checklist when buying a property.

In a chapter covering insurance and surety bonds, an expanded article is devoted to workmen's compensation insurance. You must be covered with this type of insurance if you are building your own house. Otherwise, you may be personally liable for the worker's compensation benefits due to the injured worker, and to the family and dependents in the case of the death of the worker.

Closing and closing costs are important subjects in real estate transaction. Closing, also called settlement, is the process whereby the seller delivers the agreed-upon deed to the buyer, and the buyer delivers the agreed-upon price to the seller. In most cases, the closing of the buyer's mortgage loan and the payment of the seller's mortgage, if any, takes place in the same session. In the closing proceedings, it is extremely important that the seller's mortgage be satisfied before the deed is delivered. If not, you would be buying the property with the old mortgage attached to it. Closing costs can be significant, particularly in areas where real estate taxes are high. The real es-tate settlement procedure act (**RESPA**) requires that lenders provide you with a booklet entitled *Settlement Costs*, and a good faith estimate of settlement costs before the closing date.

An important feature of any house is its architectural style, that is its look as seen from the outside. To help the reader identify the different house styles and choose the style he or she likes the most, 34 of the most popular styles in varying sizes are presented.

There are several ways to buy a house design: (1) from ready-made design books (home plans or stock plans), (2) by retaining an architect or a building designer to custom design your house, and (3) by using a house you saw either in the neighborhood or in one of the home plans books as a base line and hiring an architect, building designer, or a draftsperson, depending upon the extent of the changes, to modify the design to suit your requirements and needs.

Some people prefer to make an addition to their home rather than buying a bigger one. Planning and feasibility studies of home additions are presented in a separate chapter.

Before construction starts, you must apply for and obtain a building permit from the building department of the city or town where the property is located. The building permit application form and the documents that must be submitted with the application form are explained in detail, with special reference to zoning ordinances, building codes, energy conservation codes, staking out the house, heat sheet, soil percolation tests, well water tests, and the inspection schedule list.

Building your house successfully depends a great deal on how you go about choosing and dealing with contractors. Before construction starts, you will have to decide on whether you will hire a general contractor to build the entire house or subcontractors to build parts of the house. The advantages and disadvantages of both methods are presented so that you can make your choice based on your capabilities, and available time. How to evaluate and deal with general contractors and subcontractors is covered from both the technical and human points of view.

The next 24 chapters (19 through 42) give illustrations of the major components of construction: excavation, blasting, foundations, house frame, etc. They are written in accordance with the sequence of construction. Where more than one kind of building material can be used, such as in exterior siding or flooring, the advantages and disadvantages of each material are presented. At the end of each chapter, an article is devoted to how to contract for the item covered in the chapter. Guidelines for cost estimates and workers' productivity are provided wherever applicable. The actual costs depend on the market condition of each area as determined by competitive bidding. To help the reader make cost estimations, the labor costs of several items are given as a ratio of the material costs. When using this approach, prevailing local construction activity must be taken into consideration. In areas where construction is booming, the ratio of labor costs to material costs is much higher than that in areas where construction activity is sluggish.

Since this is your own home in which you and your family expect to live for years to come, it is recommended that you use high quality materials. This makes sense particularly since the difference in price between average and good quality materials is small. For example, concrete that has 50 percent extra strength costs only 5 percent more. The same is true for roof covering materials where the difference in price between 15- and 25-year-guarantee asphalt shingles is insignificant when compared to the total cost of roofing.

The last chapter explains the documents and steps required to obtain the certificate of occupancy (**C. of O.**), after which you can legally occupy the house. Also explained is the method by which the taxes are levied on your house and the course of action you may follow if you think that the taxes are too high.

As in any project, things can go wrong in the course of building your house. Studying this book thoroughly at the outset should help you avoid costly mistakes. To further reduce the possibility of running into snags, here are a few recommendations:

1. All the legal, design, and construction work should be done by professionals: lawyers, architects, engineers, carpenters, plumbers, electricians, tapers, plasterers, etc. If you wish to build a part of the house with your hands, study one of the how-to books on the subject you are interested in.

2. Retain an architect or engineer for construction advice and consult with him or her periodically or whenever you feel that something may go wrong.

3. Let your presence around the workers be felt without interfering with their work. Everyone will get the message—the owner is watching.

Building your house is a serious business, but it can be an enjoyable experience as well. You are in full control of selecting each item. You will have a lot of fun visiting the showrooms for plumbing fixtures, ceramic tiles, resilient tiles, wood flooring, carpeting, kitchen cabinets, vanities, windows, doors, and electrical and gas appliances. Showrooms are packed with seemingly endless varieties that dazzle the eyes, appeal to different tastes, and fit into every budget. Having the family participate in selecting each item, its style, color, and material, is a joyful and memorable experience.

Watching the project develop from a vacant lot to a liveable house is a thrilling experience. As soon as the rough framing is up, the entire family, kids and adults alike, will be anxious to visit the house and explore what there is of it. Everyone will try to figure out the layout of the house: Where is the kitchen? This is the bathroom . . . here is my bedroom . . . . Don't forget to take pictures—they will provide you with something to look back on throughout your life.

After you move into the house, you and your family will share a feeling of accomplishment and satisfaction. Your relatives and friends will be anxious to talk to you about it and learn from your experience.

Wishing you and yours the best of luck in BUILDING YOUR OWN HOME!

CHAPTER 2

Buying A Lot

A lot is a piece of land that is approved by the local building department for use as a building site for a house. In different parts of the country, a lot is called a **parcel, land, property,** or **building site.** Land is comprised of the surface of the earth and what is below it down to the center of the earth, and what is above it up to infinity, and includes all the natural resources such as trees, streams, oil, gas, and minerals. In addition to the physical quantity, land ownership includes a bundle of legal rights among which are the right to sell, lease, build, mortgage, encumber, use, enjoy, occupy, give away, share, will, mine, drill, and farm. These legal rights run with the land, meaning they benefit successive owners.

The law considers each piece of property to be unique. The legal implication of this is that one parcel cannot be substituted for another. If you bargain for a particular lot, the seller cannot make you accept a different lot.

Land with permanent man-made improvements such as a house or a barn is called **real estate** or **real property. Estate** means the extent of one's rights in real property.

Factors to be Considered

Choosing a lot involves many factors: financial, technical, educational, commercial, social, transportational, recreational, and legal, to name a few. The principal financial factor is the price of the lot, which is determined mainly by its location, size, and amount of the property taxes. Technical factors include the availability of municipal sewer and water lines; the level of the subsurface water and the water table; the grade slopes; the quality of the bearing soil; and whether the lot is located in a flood-prone area. Other concerns would be the quality of the neighborhood schools and colleges, and the proximity of shopping centers, supermarkets, social clubs, houses of worship, mass transit systems and highways, as well as theaters, beaches, parks, public swimming pools and ski lifts. Legal factors include type of ownership, type of deed, deed restrictions, easements and encroachments. (Legal terms are explained throughout the legal chapters and in the glossary at the end of the book.)

Building lots that are for sale can be found through the following means: (1) real estate agents; (2) "For Sale" signs posted along the roads and highways; (3) local newspapers; (4) friends, relatives, and co-workers; and (5) developer-builders. It is to be noted that a developer-builder does not sell an empty lot at a "reasonable" price unless he is pressed for cash, since he can make more money selling a finished house.

Purchasing the Property

There are two ways to buy a lot.

Indirectly When you contract a **developer-builder** (also called a **developer**) to build you a house on one of the lots in the **subdivision** he owns, he may show you several house designs, each with a different price, and ask you to choose one of them. In many instances, he can take you through either model homes or houses under construction that are similar to the ones he is offering.

This method involves a minimum amount of work on your part and is very convenient if you have neither the experience nor the desire to get involved in construction or the administrative work associated with building a house. The developer should either have the experience to supervise construction, or hire professionals to do this work for him. You need not worry about how much money to allocate for the lot and how much for the construction.

The disadvantage of this method is that your options in selecting your design are limited to the few models the developer has to offer. Another disadvantage is that you are required to sign a contract for the entire house at the outset without having a real chance to study the design or the materials used in construction, and before you have time to decide whether the offered design fits your requirements and needs. You may request design changes or construction alterations, but the developer is likely to discourage any deviation from the original plans since any change requires effort, time and money. He will undoubtedly charge you extra money for any change you want, but he may not have the time to do it.

Directly Buying a lot outright, you choose a house design and either hire a general contractor to build you the entire house, or assign parts of it to professional subcontractors if you feel that you are able and willing to take charge of building your house. Such an endeavor requires a considerable amount of effort, time and know-how, but the personal rewards are great. Being able to choose the design of your own house is a joy in itself. Selecting the finishing items such as bathtubs, kitchen cabinets, tiles, windows, doors, etc. is a thrilling and memorable experience for the entire family. It is financially rewarding in that you will save most of the general contractor's profit and overhead expenses. I say most, because a general contractor gets materials and labor at a discount not available to an individual building only one house. You may also deduct a portion of the points and interest on the construction loan from federal, state, and local taxes. For details, consult an accountant.

Financial Planning

At the outset you should estimate how much money to allocate for buying the lot and how much for the construction. The ratio between the price of a lot and the total cost of the house depends on the market conditions and the location. Land is cheaper in rural than in urban areas. Therefore, the price of the lot as a percentage of the total cost of the house is likely to be less in rural than in urban areas.

When comparing lot prices, keep in mind that the cash price of a lot is not the only item to be considered in the financial planning. For instance, if the lot is rocky, you will have to spend a lot of money on blasting; if it is heavily wooded, the land will have to be cleared; if the grade is steep, you will have to spend money on grading or importing fill; if there is no municipal sewer line to connect to, you will have to construct a sewage disposal system; and if there is no municipal water line near by, a well must be drilled.

You should also inquire about the amount of real estate taxes you are expected to pay since this will have an effect on the amount of the mortgage loan you can obtain to finance the house, and on your future financial obligations as well. Remember that real estate taxes are likely to increase with time.

Visual Inspection

After you find several lots that meet your requirements and financial ability, you should inspect them visually and narrow down your choices. Visual inspection of a building site reveals a lot of information. At the outset either the owner, the real estate agent, or the developer should provide you with a map or land survey indicating the location of the lot, its size, and shape. The size of the lot can be indicated either in acres or square feet. An acre is equal to 43,560 square feet.

High on the list of the items to be checked is the availability of the utilities: municipal water and sewer lines, electricity and telephone lines. If one or more of these items is not available, you should get an estimate of how much it would cost to install an alternative system.

Also check the elevation of the building site with respect to the surrounding area. If it is relatively high, it is an indication that rain water flows away from the lot, which is good. On the other hand, if it is relatively low, it indicates that rain water of the surrounding area overflows into the lot causing a wet basement, or ponding.

Subsurface water and high water table problems may not be visible at the time of inspecting the site, but might crop up either during construction or after you've built the house and moved into it. Subsurface water problems appear after a prolonged rainfall, causing a wet basement and a malfunctioning septic tank. Since it may not be raining at the time of inspection, you might ask the neighbors whether they experience a wet basement problem after heavy rainfalls. High water table may result from a nearby stream, drain, or shallow rock formations. The underground water level may be checked by digging a hole where you intend to build the house to a depth equal to that of the basement or until you reach water. If you encounter wet soil, keep the hole open for a couple of days until the water level stabilizes. If the water level is high, you know that you have to raise the level of the house, do without a basement, or look for another lot. (Precautions should be taken to ensure that nobody trips into the open hole. After serving its purpose, the hole should be backfilled.)

Visualize where you want to position the house within the lot. Ideally, the house should be built on the highest ground, where undergound water problems are at a minimum. However, the topography of the lot may indicate that this is not feasible. For example, the highest ground may be at a corner of the lot where you are not allowed to build the house. The solution to this problem is to grade the site, import fill, or install an adequate drainage system. If you think that underground water may be a problem, you should consult an engineer before starting construction.

Grade slopes should not be over 10 percent, otherwise it would be difficult to walk on, to use it as a driveway, or to maintain a lawn or any vegetation. Heavy rain can wash away the top soil taking the lawn with it. More serious, however, is when the soil is slippery (technically known as **frictionless**). It may slide down taking the house with it!

Avoid building sites that are on a cliff, a waterfront or a shoreline, where sliding or soil erosion is a possibility. Insurance companies may refuse to insure homes at these locations because they consider them to be at high risk. If you cannot get a home insurance policy for the house you are going to build for yourself and your family, you are well advised to look somewhere else.

Checking the Soil

Soil is the loose upper layer of the earth. It is important that the soil be permeable enough to allow rain water to seep through without ponding, but not so permeable that it won't retain the moisture needed for the growth of lawn and shrubs.

More important however, is the **bearing soil,** which is the layer of soil upon which the footings are poured. It must be strong enough to safely support the house throughout its life. Soil should also be uniform so that it does not exhibit differential settlement. This type of settlement, unlike even settlement, is very harmful because it causes the foundations and the walls of the house to crack.

Check whether the lot has a soil survey or topography (topo). Such a survey would be found in the office of the local soil conservation district of the county in which the lot is located. Soil surveys provide valuable information regarding the location and extent of different types of soil, as well as soil properties such as friction capability or lack of it; the existence of clay types that swell when they gain moisture and shrink when they dry; and the depth to the rocks.

If you are going to install a sewage disposal system, you must check that the soil is porous enough to absorb the effluent of the septic tank. This is determined by a **soil percolation test.** Information about this test and its procedure can be obtained from the local or state department of health.

Flood Checking

You should determine at the outset whether the lot you are interested in buying is in a **flood-plain area.** The **Department of Housing and Urban Development (HUD)** maintains maps identifying these areas throughout the country. The federal government provides a **subsidized flood damage insurance program** administered by HUD for property owners in these areas. This flood insurance is mandatory for properties financed by mortgages, loans, grants or guarantees from a federally insured or regulated lender or from a federal agency.

Some people choose to build their houses in flood-plain areas because land there is cheap. Should you decide to buy a lot in one of these areas, check with the county or state commission to see how frequent and severe the flooding is and what plans are underway or contemplated to curb flood damage effects.

Legal Checking

As a buyer, you are required by law to physically inspect the premises (the lot) for rights that others may have to the property. Such rights may include easements and encroachments.

An **easement** is a right that may be exercised by one party to use the land of another for specific purposes. An example of an easement is the right of the electric utility to use the land of an owner to lay its cables or install its poles. Another example is when an owner of a lot has a right to cross the neighboring lot to get access to a road or lake. Easements **run with the land,** and are not cancelled when ownership to either lot changes hands.

An **encroachment** occurs when a part of a building, a fence, or a driveway extends illegally beyond the land of its owner into the neighboring lot. Encroachment can be verified visually or by a land survey. If the encroachment is less than 10 years old, the encroached upon property owner may sue for removal of the encroachment or seek damages. But, if the encroachment is more than 10 years old, it may become an easement and so become a permanent right.

In addition to physical checking of the lot, you should ensure that the lot is not earmarked to be seized for public use. A federal or state government or a public service organization can acquire a privately owned lot through a court action called **condemnation.** The statute under which this power is given is called **Eminent Domain.** The law requires that the owner of the seized property must be fairly compensated.

Also check the type of deed you would be getting because all deeds are not alike. You are likely to get a bargain and sale deed (see Chapter 7).

Types of Ownership

There are four types of property ownership: (1) Fee Simple or Fee Simple

Absolute, (2) Fee Simple Determinable or Qualified, (3) Fee Simple on Condition, and (4) Life Estate.

1. **Fee Simple or Fee Simple Absolute.** This is the highest form of ownership. It entitles the title holder of a property to unconditional power of disposition during the person's life and descending to the person's distributees and legal representatives upon his or her death.

2. **Fee Simple Determinable or Qualified.** This can easily be identified by the language of condition such as the words "as long as," "while," and "until." For example, a grantor (seller) conveys a property to a grantee (buyer) as long as he does not open a liquor store on it. If the grantee opens a liquor store on the property, ownership goes back **automatically** to the grantor.

3. **Fee Simple on Condition.** Its language is different from the fee simple determinable in that it states a condition that if not fulfilled, the grantor has the right to repossess the property. For example: "To the grantee, but if the property is not used for recreational purposes, the grantor has the right to reenter and repossess the property." In this type of ownership, the right to repossess is not automatic as in the case of the fee simple determinable.

4. **Life Estate.** Life estate ownership is limited in duration to the life of the grantor, grantee or a third party. For example, a grantor grants a property to a grantee for as long as the grantee lives; after the grantee's death, the property is granted to a particular charity. The grantee can sell the property to a buyer. The buyer will have absolute ownership (fee simple) to the property as long as the grantee lives. When the grantee dies, ownership goes to charity. The duration of life estate may not necessarily be tied to the life of a human being. For example, a grantor grants a property to a grantee as long as the grantor's beloved dog, Rex lives; after Rex's death, the property returns to the grantor's heirs. The grantee has absolute ownership (fee simple) to the property as long as Rex lives. Should the grantee die before Rex, his heirs inherit the property for the duration of Rex's life.

You should be certain that you buy a property that has a **fee simple** type of ownership.

Touring the Area and Talking to the Neighbors

Touring the area around the building site gives you a very good chance to get a sense of the neighborhood. You will notice the size and style of the surrounding houses, and how they are maintained and landscaped, and discover the location of the nearest shopping centers, schools, houses of worship, and roads and highways.

While touring the area, try to talk with the residents. They can provide you with valuable information regarding local conditions. You will find them for the most part eager to talk to you since they are curious to know who is going to be their future neighbor.

When my wife and I were searching for a lot, a real estate agent showed us one in a beautiful neighborhood. However, it was underpriced, which made me suspicious that the lot might have a subsurface water problem. When I asked the real estate agent if this was the case, she said she did not know. Next day, my wife and I went alone (without the agent) to see the lot and tour the area. We found a neighbor mowing his lawn. I greeted him and we started talking. I found him very curious to know whether I was going to buy the lot, how much was the asking price, where did I live and what do I do for a living, does my wife work and do we have children. On my part, I was interested to know whether the subsurface water or water table was high enough to cause wet basement problems. Because it requires effort, time, and money, it is simply not practical to dig a hole in every lot we look at to determine the level of the underground water. In most cases, asking the neighbors can provide the correct answer. It is also scientific because it is based on experience.

My conversation with the neighbor confirmed my fears about the subsurface water. He told me that the house adjacent to that lot had serious basement water problems and that its owner was pumping water all the time. The neighbor then called his wife and introduced us, invited us inside their home, and coffee and cake followed.

During our visit, they told us about two lots that were owned by people who had moved to the west coast and who were willing to sell. We saw the lots and found them to be on high ground, an indication that underground water would not be a problem. Our host called the lots' owners on our behalf to find out their asking price. The price was reasonable but my wife did not like the lots, and we had to look elsewhere.

There are two points to be made here: (1) Talking with the neighbors can provide you with valuable information regarding local conditions and services, and (2) the neighbors may know someone who is selling land.

After you satisfy yourself and your family that a particular lot is the best you can get whether or not it includes a finished house, you will have to enter into a **Contract of Sale** with either the developer or the lot's owner.

CHAPTER 3

Contracts and Arbitration

A **contract** is an agreement between competent (sane adult) parties to do certain things that are legally enforceable. In the course of building your house, you will enter into a **contract of sale of real estate** either with a developer to buy a house including the lot, or a landowner to buy an undeveloped lot. For the latter case, you will enter into additional construction contracts with either a general contractor to build the entire house or several subcontractors, each to build a specific part of the house.

Essentials of a Contract of Sale of Real Estate

The contract of sale of real estate must explicitly contain all the terms and conditions agreed upon between you (the buyer) and the developer or lot owner (the seller). Implied agreements or understandings are not enforceable. For example, if the seller does not explicitly promise to deliver a full covenant and warranty deed, you may be required to accept a bargain and sale deed.

The essentials of a valid contract of sale are:

1. There must be a **meeting of the minds,** meaning a mutual agreement between buyer and seller on the provisions of the contract.

2. The contract must be in writing and signed by both parties. (This is required by the **Statute of Frauds** in all the states.)

3. The parties must be identified.

4. The parties must be competent.

5. An expression of the seller's agreement to sell and the buyer's agreement to buy.

6. The property must be sufficiently described.

7. The price (valuable consideration) and the terms of payment.

8. Agreement of the seller to convey title. The contract usually specifies the type of deed to be delivered to the buyer, although this is not essential.

Encumbrances

An **encumbrance** is a right to the property by a third party. It often limits land use but does not prevent transfer of title. Where there are encumbrances in the title to be conveyed, the contract must express the agreement of both parties with respect to them. Some of the most common encumbrances are:

1. Unpaid taxes and assessments for local improvements. They automatically become liens (see Chapter 5) to the property.

2. Mortgages.

3. Lease of the property or any part of it.

4. Judgments against the seller, duly recorded in the county in which the property is located.

5. Mechanic's liens for work performed or material furnished for use to improve the value of the property.

6. A legal document (*lis pendens*) filed in the office of the county clerk, giving notice that an action or proceeding is pending in the courts that may affect the title to the property.

7. Encroachment of a structure on the land of the adjoining owner or a street.

8. Easement (the right of one party to use the land of another for specific purposes).

9. Deed restrictions.

10. Code or zoning violations (discovered by a recent land survey or certificate of occupancy).

Time is of the Essence

This clause is used if either the buyer or the seller has compelling reasons for closing the title upon a fixed date stated in the contract. By that date, the seller must be able to deliver the deed to the property and the buyer must be able to pay the agreed upon price. If either party is unprepared by then to fulfill its obligation, it may suffer substantial financial losses. Thus, extreme caution should be exercised when including this clause in the contract.

Offer to Purchase

In many parts of the country, buying real property begins with an **offer to purchase,** which is sometimes called a **binder** or **earnest money agreement.** It is an informal agreement, usually drawn by a real estate agent, in which you (the buyer) set your terms and sign it. The agent may require from you a small cash deposit, for which he gives you a receipt, as evidence of your "good faith" to complete the transaction. Subsequently, the agent presents your offer to the seller. If the seller accepts it, a formal contract of sale is drawn within a few days. Care should be taken to ensure that the terms of the final contract are exactly the same as those of the offer to purchase.

You, the buyer, may withdraw the offer to purchase at any time prior to its acceptance by the seller. In order not to tie up your offer for a long period of time, you are advised to specify a deadline for its acceptance, beyond which time your offer expires automatically. If you withdraw your offer before it has been accepted or if your offer is rejected or has expired, you are entitled to the return of all your deposits.

Shopping for Services

When the seller accepts your offer to purchase, you should start shopping for services such as a mortgage, lawyer's services, and a title insurance company. Once you've secured a mortgage commitment and retained a lawyer, you should convey this information to the seller either directly or through the real estate broker, if there is one. Subsequently, the seller asks his or her lawyer to prepare a formal contract of sale. A contract is then drawn, signed by the seller and mailed to your lawyer for his review and to obtain your signature and the deposit check. At this point, your lawyer advises you as to what your losses will be if you default, based on the terms of the contract. Next, your lawyer mails both the signed contract and the deposit check to the seller's lawyer who keeps them until the closing. If you do not know of a title insurance company, your lawyer may recommend one to you.

Deposits

When you and the seller sign the contract of sale, it is common but not essential that you enclose a cash deposit with the contract. You should not pay a deposit, or sign a contract, unless you are fairly certain of your ability to complete the transaction, since the deposit may be retained by the seller if you default, depending on the terms of the contract.

The amount of the deposit is subject to negotiations between the seller and the buyer. Naturally, the seller likes to get a large deposit to compensate him for any losses he may incur if you default and to pay for his lawyer and the real estate brokers. On your part, you like to pay a small deposit to minimize your losses should you become unable to complete the transaction. In most contracts, the deposit is equal to 10 percent of the purchase price.

This deposit may be held by either the seller's lawyer, or the real estate broker in a special escrow account. The escrow holder may maintain one escrow account for deposits from all her clients, but she is not permitted to commingle the escrow money with her personal funds, and she must maintain complete and accurate records of all her customers' deposits.

Real Estate Commission

The real estate commission is a fee for services. It is usually paid by whomever hires the broker or agent. In most cases, it is the seller. (The party that hires the real estate agent is called the **Client** or the **Principal,** and the other party is called the **Customer.**) A buyer may pay the real estate commission provided that this fact be known to all parties, and that they agree to it. Also, a buyer may hire a broker to find him a parcel. In this case, the buyer usually pays the broker's commission; the broker must make this fact known to all parties.

There is no rule for how much the commission should be. The client may either negotiate the commission with the broker or shop around for better rates. For land sale, it varies from 4% to 10%, but in most cases, the range is 5% to 7%.

The broker who obtains the **listing** from the seller is called the **listing broker.** Listing is an agreement between the seller and the real estate broker by which the broker is authorized to represent the seller in soliciting offers to buy the property in return for a negotiated fee payable when the property is sold, usually at the closing.

The listing broker may give notice to

other brokers that he has such a property for sale in case one of them has an interested buyer. This is widely done through a **Multiple Listing Service (MLS).** This service is organized by a group of brokers within a geographical area to make any member's listing available to the other members. This gives wide exposure to properties listed with any MLS member.

If a broker other than the listing broker finds a prospective buyer, she has to present the offer to purchase to the listing broker who must submit it to the owner. If the offer is accepted and the property is sold, the broker who found the buyer is called the **selling broker.** In this case, the fee or commission is usually, but not always, split between the two brokers. If for example the fee is 6 percent, each broker gets 3 percent.

Further, a broker usually hires sales representatives to work on her behalf on a percentage basis that is negotiated between both.

The listing and selling brokers and their sales representatives owe their loyalty to their employer who pays them, usually the seller, and they must not disclose whether the seller will accept a price lower than the listing price or any facts that might impair the seller's bargaining position. They try to get the seller the best and highest offer. The highest offer is not always the best because it may be accompanied by terms unacceptable to the seller such as the seller being required to finance the deal, or the buyer offering a very small deposit or requesting a specific closing date that the seller cannot meet. Thus, you should not confide in a real estate broker or representative regarding the offering price or your financial posture. This is not because she is not

trustworthy, but because she represents the seller. She is not supposed to counsel you. For example, if you tell her: "I will offer $40,000 for the lot, but I can go up to $45,000, what do you think?" you would be putting her in an awkward position, because it is her duty and in her interest to get the seller the best and highest offer. In this case, she is obliged to convince you to offer the $45,000 immediately. Obviously, she will make a bigger commission if you buy the lot for $45,000.

Default

A **default** is a breach of contract. Specifically, it is the failure by either the seller or the buyer to fulfill a promise or meet an obligation stated in the contract of sale agreement, such as when the seller refuses to convey title or the buyer is unable to pay the price. It is prudent for both parties to include in the contract what the penalties will be if either the seller or buyer defaults. Unless otherwise stated in the contract of sale, the following are possible legal remedies if either party defaults.

If the seller defaults, the buyer may:

1. Rescind (cancel) the contract and recover his deposit and all expenses he incurred such as fees for a land survey or mortgage application.

2. File a lawsuit called **Action of Specific Performance** to compel the seller to convey the title. The basis for this lawsuit is that real property is unique and the property the seller agreed to sell cannot be replaced.

3. Sue the seller for damages.

If the buyer defaults, the seller may:

1. Declare the contract forfeited and retain the buyer's deposit.

2. Sue the buyer for damages.

3. Rescind the contract and return the deposit to the buyer. This happens when the seller finds another buyer who is ready, able and willing to buy the property for a price equal to or more than the price in the breached contract.

Printed Contract of Sale

Laws do not require that a contract of sale of real estate be written in any specific form. In practice, however, most contracts of sale are executed on standard printed forms with the parties to the contract or their lawyers deleting or adding some clauses (the added clauses are called **riders**) to reflect their mutual agreements. A typical printed contract of sale includes the following:

1. The date of the agreement.

2. The name and address of the seller.

3. The name and address of the buyer.

4. A description of the property. (A more detailed description may be included in a separate document designated **"Schedule A."**)

5. All fixtures included in the sale. (For a new house, the stove is the only appliance that is required to obtain the certificate of occupancy. Other items such as the refrigerator, washing machine, clothes dryers, freezers, wall to wall carpeting, flooring, drapes, blinds, or mail box are optional and the contract must state which of these items are included in the sale.) This clause is omitted if the contract is for the sale of a lot.

6. The purchase price (valuable consideration) and the methods by which it is to be paid. Specifically, the contract of sale recites the following methods:
   a. The deposit in the form of a check subject to collection upon signing the contract.
   b. By allowance for the principal amount still unpaid on existing mortgages (if any).
   c. By purchase money note and mortgage from purchaser to seller (if any).
   d. The amount of the balance at the closing.

7. Acceptable methods of payment:
   a. Cash, but not over $1000.00
   b. Certified personal check (this is the most common method) or official bank check, from a savings bank, savings and loan association or trust company, payable to the order of the seller or any other party the seller or his lawyer designates, such as the mortgagee of the property.

8. A clause under the heading "Subject to Provisions." (This is a very important clause because it includes riders identifying the easements, deed restrictions and so forth.) It states that the premises are to be transferred subject to:
   a. Laws and governmental regulations.
   b. Restrictions and reservations made by the seller in a declaration executed on . . . and recorded in the office of . . . on . . . in Liber (recording book) . . . page . . . of Deeds. (The blanks are filled in if deed restrictions are included in a separate declaration.)
   c. All other restrictions, covenants or easements as shown in the public records or a land survey. (A typical easement is the one granted to the electric utility to use the property to lay its cables or erect its posts or transmission lines.)

9. A clause explaining the meaning of the word **closing,** and the type of deed to be delivered. The seller then promises that the deed is in proper statutory

form for recording so as to transfer full ownership (*fee simple title*) to the premises free of all encumbrances except as herein stated. . . . (All encumbrances should be stated here.)

10. The place and date of closing.

11. The seller's agreement to pay the real estate broker's fee or commission, and the purchaser's promise that he did not deal with any broker other than that named in the contract.

12. The seller's agreement to sell his ownership and rights, if any, to any land lying in the bed of any streets or highways opened or proposed in front of or adjoining the premises.

13. The seller's agreement to deliver at the closing a certificate, dated not more than 30 days before closing and signed by the holder of each existing mortgage, stating the amount of the unpaid principal plus interest, date of maturity, and interest rate. (This item is required if the existing mortgage is to be assumed by the purchaser, and is omitted if the purchase price is used to satisfy all outstanding mortgages, or if there is no existing mortgage.)

14. The seller's promise that he has not violated any governmental or municipal laws affecting the premises and that the premises shall be transferred free of such violations at the closing.

15. A clause stating that all money paid to the account of this contract (including the deposit), and the reasonable expenses of title examination, land survey, and house inspection are made liens on the premises and are collectable out of the premises. Such liens are cancelled if the buyer defaults.

16. A clause stating that if seller is unable to transfer title to purchaser, seller's sole liability shall be to refund all money paid plus all expenses incurred for examining the title, additional searches, survey, and survey inspection. Upon such refund and payment the contract shall be cancelled. (This means that you cannot sue the seller if he defaults. If you want to have the right to sue him, you have to omit this clause and both you and the seller must initial it.)

17. A clause stating that the purchaser has inspected the building on the premises and all the items included in the sale. (This item is omitted in the sale of an undeveloped lot.)

18. A clause stating that all prior understandings and agreements between seller and purchaser are merged in this contract. (This means that this contract **overrides** the offer to purchase, or the binder agreement.)

19. A clause stating that the contract may not be changed or cancelled except in writing.

20. Several clauses stating how taxes, assessments, water meter readings, and oil already in the tank (dollar value) are to be divided between seller and buyer.

21. A rider naming the real estate listing and selling brokers and the amount of their commission, and stating that the commission is to be divided between them as they agree.

Contracting a Developer

Your contract with a developer is basically a contract of sale of real property to which riders pertinent to new construction are added. Such riders include:

1. Description of the house either by referring to a particular standard model or a set of drawings and the corresponding **specifications**.

2. Any options or particular specifications other than those of the standard model.

3. A clause stating the expected dates of obtaining the certificate of occupancy and the closing.

4. A clause stating for how long the developer guarantees the construction (at least for one year after the house is occupied).

The agreement may include other provisions to suit the conditions of the particular lot, such as "if rocks are encountered, the buyer may get a smaller or no basement, otherwise, the buyer has to pay the costs of blasting and trucking the rocks away." Obviously, the developer cannot give you an estimate for blasting costs in advance because he does not yet know the extent of the rocks he may encounter. Nevertheless, you should negotiate this point with the developer in detail since it can be a loophole for him. For instance, if you cannot afford blasting, he may reimburse you little or nothing at all for not building the basement. And if you decide to blast, he may charge you a much higher price than what it costs him (you would not know how much the real cost is), and may not credit you with the savings he realizes from not having to do soil excavation. Furthermore, in order to increase his profit, he may truck the rocks away at a considerable cost instead of burying them on-site at little or no cost to you.

Once you have agreed upon all items including the price, the developer may want to determine whether you qualify financially to buy the house. Hence, he draws up an "offer to purchase" agreement since it is doubtful that you would have a mortgage commitment by that time. He may even recommend a particular lender to you to expedite

your mortgage application. Before you sign a contract, add a clause stating that this agreement is subject to your lawyer's review within a specific number of days. This gives you the opportunity to review the contract with your lawyer, your family, and friends. Another precaution is to add a clause that this agreement hinges upon your ability to obtain adequate financing, otherwise the agreement is cancelled without any cost to you, and that you will get all your deposits or good faith money back.

Contracting a Lot Owner

Contracting a lot owner is different from contracting a developer in that it involves the process of **offer and acceptance** negotiations. It is similar to buying a previously owned house.

It starts with an owner offering a lot for sale, either directly or through a real estate broker. He sets his asking price and other terms such as the closing date and method of payment. If you are interested in buying the lot, you call the seller or his broker to learn if the lot is still on the market. If it is, you make an offer in writing in which you either accept the seller's terms or set your own. In some parts of the country, this offer is considered a precontract agreement or a binder, meaning that it is not binding to you even if the seller accepts your offer and signs it. The formal contract will be prepared by the seller's lawyer after you and the seller agree on all the terms. In other parts of the country the offer is in the form of a contract and becomes enforceable when the seller accepts and signs it. Thus, you must be aware of the legal status of your offer.

If the seller changes any terms in

your offer, he creates a **counteroffer.**
This relieves you of your original offer
because the seller has in fact rejected
it. The seller's counteroffer is presented
to you and you either accept it or counteroffer. This procedure continues until
either you or the seller agree to the other's latest counteroffer. An offer is considered accepted when the party
making the latest offer is notified that
the other party has accepted his offer.

Contracting a General Contractor

After you have bought a lot and se-
lected a house design, you may decide
to assign the construction of the entire
house to a **general contractor.** Usually,
you should solicit bids from at least
three contractors, preferably on **lump
sum** or **fixed price** rather than **cost plus**
basis. The latter involves a lot of paper
work, with potentials for disagree-
ments. Either way, you should draw
up a contract with the one you choose.
The contract should include the following:

1. The date of the agreement.

2. The name and address of the owner
   (yourself).

3. The name and address of the general
   contractor.

4. Location of the construction site (lot).

5. At least one set of the detailed drawings (**blueprints**), and specifications.

6. Date of start of work and date of completion.

7. The name of the construction super-
   visor, if any.

8. The price of construction either as a
   fixed price or the contractor's costs
   plus a percentage to cover his profit
   and overhead expenses.

9. Method of payment (usually in installments to be paid during the course of
   construction).

10. A clause stating that the price includes
    labor, materials, tools, scaffoldings,
    water, electricity, heat and any other
    services needed to complete the work.

11. A clause stating that the contractor is
    responsible to obtain the permits and
    the approval of the town inspectors for
    all stages of construction.

12. A clause stating that the contractor
    shall clean the house and the site upon
    completion of work.

13. A clause describing the lawn and land-
    scaping that are included in the con-
    tract.

14. A clause stating that the contractor is
    legally responsible for the safety of his
    workers and passersby and that the
    owner is not liable in case of their in-
    jury or death.

15. A clause stating that the contractor
    shall present to the owner a proof that
    he carries a workmen's compensation
    insurance policy.

16. A clause stating that the contractor is
    liable for a penalty of $ _____ for each
    day's delay. Additionally, if the con-
    tractor defaults, or stops working be-
    fore his work is completed, the owner
    reserves the right to hire another con-
    tractor to complete the work at the
    original contractor's expense seven
    days after he is given a notice to re-
    sume work by certified mail.

17. A clause stating that extra costs re-
    sulting from unforseen conditions
    (such as rocks or poor soil), any
    changes in the drawings, or any ad-
    ditional work to be performed shall be
    negotiated between the owner and
    contractor.

18. A clause stating that all unresolved
    claims and disputes shall be resolved

by arbitration according to the rules of the **American Arbitration Association.**

19. A clause stating that the contractor guarantees his work for at least one year after the date of obtaining the certificate of occupancy, and that he promises to fix at his expense any defects that may crop up during the warranty period, whether these defects are structural (such as cracked walls or foundations) or operational (such as doors or windows that do not close properly or an inoperable heating system).

20. A clause stating that no liens on your house be filed by him or any of his subcontractors or material suppliers.

Contracting a Subcontractor

You may decide to directly subcontract different parts of the house to professional subcontractors. One method is to buy the materials and contract labor only. This can be done for rough framing, tiles, roofing, windows and doors. This method allows you to control the quality of the materials and it may save you some money, too. But it requires vigilant effort on your part. The other way is to contract both materials and labor. This method is recommended for plumbing because the fixtures are fragile and it would be difficult to determine who is responsible if a piece breaks or cracks; and for rough electrical since the material costs are very small compared to labor.

Whatever method you choose, you should solicit bids or proposals from at least three contractors. Based on the drawings and specifications, each one will submit to you a proposal including the work to be done, brief specifications, the price, and method of pay-

ment. Depending on the rapport between you and the subcontractor, and the amount of work involved, a subcontractor may submit his proposal to you verbally, on the back of his business card, or on a printed form in which he fills the blanks. Neither of you is required to sign it in order to create an enforceable contract.

Generally, a subcontractor likes to have a written contract, particularly if he has never dealt with you before. A printed form is most convenient for him. Read it in its entirety before signing since these forms are not all alike. A typical form includes the following:

1. All work is to be completed in a workmanlike manner according to standard practices.

2. All material is guaranteed to be as specified.

3. Any alteration or deviation from above specifications involving extra costs will be executed only upon written orders, and will become an extra charge over and above the estimate.

4. All agreements are contingent upon strikes, accidents, or delays beyond the contractor's control.

5. Contractor's workers are fully covered by workmen's compensation insurance.

You may agree to the terms of the proposal as is, or negotiate new terms with the subcontractor until you come up with an agreement. At the bottom of the printed form there is a section entitled "Acceptance of Proposal." It runs as follows:

The above prices, specifications and conditions are satisfactory and are hereby accepted. The contractor is authorized to do the work as specified. Payment will be made as outlined above.

When you sign on the dotted line, the proposal becomes an enforceable contract.

For your protection, you may add the following terms on a separate sheet of paper that must be signed by both you and the contractor and is referred to in the proposal as follows: "The attached terms and conditions constitute a part of this contract."

1. The contractor shall in no way damage the work of others in the course of performing his work. Should this happen either intentionally or unintentionally, he must repair the damage at his own expense.

2. The contractor shall obtain the required permits and licenses needed to perform his work from the authorities and shall comply with all the rules and regulations of the federal, state and local governments (this is specifically intended for the plumber and the electrician).

3. The contractor is responsible to pay his workers, helpers, subcontractors and material suppliers, without those persons having any claim against the owner.

4. The contractor shall guarantee his work for at least one year after the house is occupied.

5. The contractor shall complete his work by _____ . If the contractor defaults or stops working before his work is completed, the owner has the right to hire another contractor to complete the work at original contractor's expense seven days after he is given a notice by certified mail to resume work.

6. No liens on your house are to be filed by him, his subcontractors or material suppliers.

Some printed contracts include a clause stating that you are liable to pay for the contractor's legal expenses should a lawsuit over the contract arise. You may delete this clause. Also try to avoid printed contract forms that are written in complicated legal jargon. Some of the language can be hard to understand.

Arbitration through the American Arbitration Association

As mentioned previously, your contract with a general contractor should include a clause stating that unresolved disputes shall be resolved by arbitration according to the rules of the American Arbitration Association. This association does not provide legal advice or arbitrate between the parties. Rather, it establishes the rules for arbitration and, for a fee, provides a list of arbitrators. It is to be noted that the laws in many states allow either party to refuse arbitration even though he or she might have agreed to it in the terms of the contract.

The procedure starts with a written request from the party wishing arbitration to the other party stating the subject of the dispute and the desire to arbitrate. If the other party agrees, arbitration is conducted by a board comprising three members to be chosen by the disputing parties from a list provided by the Association. Arbitrators should be impartial and knowledgeable in the field of construction. Then, a date is set for an arbitration hearing. Each party may be represented by a lawyer and may call witnesses to support his or her argument. Each party has the right to question the other party and any witnesses.

A few weeks after the hearing is completed, the board renders a written decision and sends a copy of it to each party. The parties are bound by the board's decision, which is enforceable by law. The decision cannot be appealed unless one party can submit a proof of bias or misconduct on the part of the board.

Court Arbitration

Some states provide court arbitration for disputes arising from construction work if the disputed amount is within a certain limit (about $6000). This type of arbitration is usually sought by a contractor who does not have a written contract or a material supplier who does not have receipts to substantiate his claim. The idea is that a contractor or material supplier must be compen-

sated for the services he rendered whether or not he can substantiate his claim with written documents. This type of arbitration is obliging to the homeowner.

In court arbitration, a board comprising three members is appointed by a judge. They must be impartial and at least one member should have knowledge of construction in addition to legal knowledge. A date is set for the hearing during which each side presents his or her case. Within a couple of weeks after the hearing, the board renders its award based on the evidence and arguments that each party has presented in the hearing. Because the disputed amounts are small, they do not warrant an appeal.

CHAPTER 4

Property Description and Land Survey

*P*roperty description is a means of positively identifying a piece of property or a lot. It is one of the essentials of a valid deed. If a property is not sufficiently described in the wording of the deed, the buyer and the seller may disagree on the exact area of the conveyed property. This can lead to lengthy and costly lawsuits and may invalidate the deed. A description is considered legally sufficient if it enables a land surveyor to locate the property. Because of its significance in conveying property, a description should be prepared only by a surveyor or a lawyer. Property description must include the name of the county and state in which the property is located.

It is possible to describe a property because each piece of land is unique at least as far as its geographic location is concerned. It is important to use the same description as that of previous conveyances to eliminate doubt and confusion. Furthermore, the description of the property in mortgages and other documents should be identical to that provided in the wording of the deed.

A property may be described by one or more of the following methods: (1) metes-and-bounds, (2) lot and block numbers, and (3) government survey system.

Metes-and-Bounds

Metes-and-bounds is an old method of describing land by specifying the measures, directions, and shapes of the property lines, and by referring to **monuments,** if there are any. The description starts at a **point of beginning (POB),** which must be a permanent well-referenced corner of the property. The description then proceeds by indicating the length and direction of each property line to the following corner, and continues until the POB is reached, thus closing the area. Property lines and corners may be identified by monuments in addition to measurements. If the actual distance to a monument is different from the measure set forth in the wording of the description, the actual distance takes precedence. Therefore, the measured values are often accompanied by the words "more or less."

For better identification, the property description may include the area of the property and the names of the adjoining property owners.

The units of measuring the lengths and directions of the property lines are explained in the following section.

Linear Measurements

Linear measurement of each boundary line is given in feet and decimals (e.g., 235.56 feet). In old deeds, measurement used to be in poles, rods, or chains. The U.S. linear measurement units and their abbreviations are:

1 foot (ft) = 12 inches (in); 1' = 12"

1 rod = 1 pole = 16.5 feet (ft)

1 engineer's chain = 100 feet = 100 links; 1 foot = 1 link (lk).

1 Gunter's chain (ch) = 66 feet = 100 links; 0.66 foot = 1 link

1 vara = about 33 inches (in)

Vara is an old Spanish measure and is often encountered in the southern and southwestern states.

1 fathom = 6 feet

1 mile = 5280 feet

The measurement units are currently being converted to the **International System, SI** (*Système International d'unités*) in many nations including the United States. In the SI system, the basic unit of linear measurement is the **meter.**

It was originally defined as equal to 1/10,000,000 of the earth's meridional quadrant. A more tangible meter measure was established as the distance between two marks on a bar stored near Paris, France. The bar is composed of 90 percent platinum and 10 percent iridium, and is known as the **International Meter.** A copy of the bar is kept at the **U.S. Bureau of Standards** in Washington, D.C.

In the SI system, conversion from one unit to another is accomplished by decimals. The basic SI linear measurement units and their abbreviations are:

1 kilometer (km) = 1000 meter (m)

1 decimeter (dm) = 0.1 meter

1 centimeter (cm) = 0.01 meter

1 millimeter (mm) = 0.001 meter

The U.S. units can be converted to SI units and vice-versa according to the following relations:

1 U.S. inch = 2.54 centimeter = 0.0254 meter

1 U.S. foot = 30.48 centimeter = 0.3048 meter

1 meter = 39.37 inches

1 meter = 3.2808 feet

Area Measurements

The units of area measurement in both the U.S. and SI systems are:

1 acre = 43,560 square feet

1 acre = 160 square rods

1 acre = 10 square Gunter's chains

1 square mile = 640 acres

1 hectare = 10,000 square meter

1 hectare = 2.471 acres

1 acre = 0.4047 hectare

Angle Measurements

The units in measuring angles are the degree, minute, and second.

1 circle = 360 degrees = 360°

1 degree = 60 minutes = 60′

1 minute = 60 seconds = 60″

To write the value of an angle, the degrees are written first, followed by the minutes and the seconds with spacings separating them. As an illustration, an angle equal to 50 degrees plus 41 minutes plus 17 seconds is written as follows:

$$50° \quad 41′ \quad 17″$$

The amount of one second is very small and could not be visualized by the naked eye or measured by a regular protractor. However, surveying instruments such as **theodolites** or **transits,** have the capability of measuring seconds. In long lines, one second makes a difference in the direction of a line and thus in the total area of a property.

Direction of Property Lines

The direction of property lines can be defined by: (1) a reference line the direction of which is well established, and (2) the magnitude and direction of the angle between the reference line and the property lines.

Reference Line   The **reference line** or **reference meridian** may be any line the direction of which is identified.

The most common reference meridians are:

1. The **true or astronomic meridian,** which is the north–south direction of a line passing through the earth's geographic poles. The true meridian is the most common reference line in land surveying.

2. The **magnetic meridian,** which is the direction taken by a freely suspended magnetic needle under the effect of the magnetic field of the earth only.

The Angle between the Property and the Reference Lines

There are two established methods of defining the angle between the property and the reference lines: (1) bearings, and (2) azimuths.

1. **Bearings.** The bearing of a line is the quadrant in which the line falls (NE, SE, SW and NW) and the horizontal acute angle (less than 90°) that the line makes with the reference meridian in that quarter. As an illustration, the bearings of the lines OA, OB, OC and OD shown in Figure 4.1 are North 60° East,

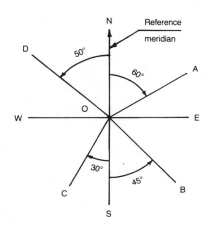

Figure 4.1 Defining the direction of property lines by the bearing method.

South 45° East, South 30° West, and North 50° West, respectively. On survey maps, these bearings are abbreviated as N 60° E, S 45° E, S 30° W, and N 50° W, respectively. The point O in this figure represents the corner at which the person who determines the direction of the line would be standing. There are two bearing classifications:

a. **True bearing** for which the reference line is the true meridian.

b. **Magnetic bearing** for which the reference line is the magnetic meridian.

2. **Azimuths.** The azimuth of a line is defined by the angle that the line makes with the reference meridian measured in a clockwise direction generally from the north branch of the meridian. The azimuths of lines OF, OG, OH and OJ shown in Figure 4.2 are 65°, 130°, 205° and 325°, respectively. The quadrants are not defined in this system.

Monuments

Monuments are fixed marks or objects that can be easily identified. They may be either natural or man-made. Natural objects include rivers, streams, large trees, stones, or road intersections. Some monuments such as rivers or roads may be used as property lines. Man-made objects include bench marks placed by land surveyors, buildings, or fences. For positive identification of property lines, some states require that all property corners be marked by monuments comprised of steel sections or concrete piers having their top above the ground and their bottom dug into the ground a few feet down or to below the frost line.

Example of Metes-and-Bounds Description

As an illustration of a metes-and-bounds land description, lot 20 on the map shown in Figure 4.3 reads as follows:

All that certain plot, piece or parcel of land, situated, lying and being in the town of Mamaroneck, county of Westchester and state of New York. **BEGINNING** *at a point on the westerly side of Carol Lane between said lot 20 and lot 21 which point is also distant South 14° 56' 49" East 132.76 feet from the southerly end of a curve having a radius of 25.0 feet and an arc length of 49.40 feet, connecting the said westerly side of Carol Lane with the southerly side of Gate House Lane; running thence from said point of beginning along Carol Lane South 14° 56' 49" East 136.93 feet to the division line between lot 20 and lot 19 as shown on said map; running thence along said division South 75' 00' 13" West 146.00 feet, running thence North 14° 56' 49" West 137.05 feet to the aforesaid division line of lots 20 and 21; and running thence along the same North 75° 03' 11" East 146.00 feet to the point and place of beginning.*

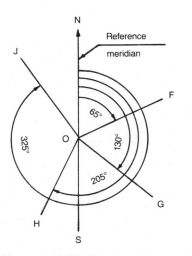

Figure 4.2 Defining the direction of property lines by the Azimuth method.

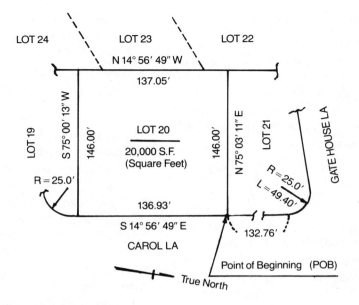

Figure 4.3 Land survey of Lot No. 20.

Lot and Block Numbers

With this method, the land is described by referring to a **tax map** or a **plat of subdivision** filed in the public recording office of the county or city where the property or lot is located. As an illustration, lot 20 previously described by metes-and-bounds may be described by lot and block numbers as follows:

All that certain plot, piece or parcel of land lying and being in the town of Mamaroneck, county of Westchester and state of New York, shown and designated as lot 20 on a certain map entitled Subdivision Map of Saxon Glen, *Section One in the town of Mamaroneck, Westchester County, New York, made by Doe Jones, dated November 21, 1965, and filed in the County Clerk's office, Division of Land Records, Westchester County, New York on June 15, 1966 as Map No. 15765.*

A part of this map is shown in Figure 4.4.

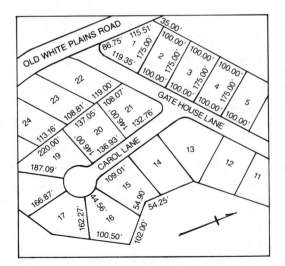

Figure 4.4 Subdivision Map No. 15765.

The Government Survey System

The **government survey system,** also called **U.S. System of Public Land Survey,** or **rectangular survey system,** was established in 1785 and is used as a standard method of describing land in the 30 states that include public land. The other 20 states in which this system is not used are the 13 original states, Texas, Tennessee, Kentucky, West Virginia, Vermont, Maine, and Hawaii.

In this system the land is divided into grids and subgrids that are approximately square in shape. One can see these grids clearly when flying over the states that use this system. The units and sizes of these grids and subgrids are:

1. **Tracts** or **quadrangles.** Each tract is approximately 24 miles square.

2. **Townships.** Each tract is divided into 16 townships. Each township is approximately 6 miles square.

3. **Sections.** Each tract is divided into 36 sections. Each section is approximately 1 square mile.

4. **Subdivision.** Each section may be divided locally into subsections of various sizes.

The reason for using the word "approximately" in describing the shape of the grids is due to the spherical shape of the earth. The true meridian lines converge as they extend north until they meet at the North Pole. As a result, the northern edge of a tract is less than 24 miles, as illustrated in Figure 4.5.

The basic elements of the government survey system are:

Initial Point   The **initial point** is an established secured mark used as a reference for the survey of a large designated area. The 30 states surveyed by this system have 37 initial points of which 5 points are in the state of Alaska.

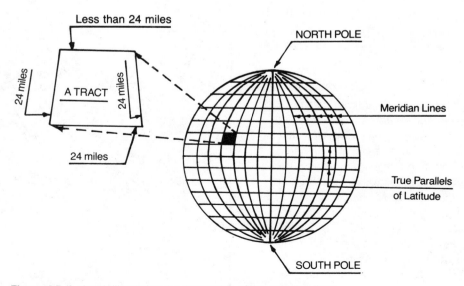

Figure 4.5 A sketch illustrating the conversion of the meridian lines.

Principal Meridian The **principal meridian (PM)** is a north–south true line that passes through the initial point (see Figure 4.6). The PM ends north and south at the boundary lines of the area that is using the initial point as a reference. The principal meridian of each area is given a unique name or number that must be referred to in all subdivision surveys.

Base Line The **base line** is a true parallel of latitude (east–west direction) that passes through the initial point. It intersects the principal meridian at a 90° angle (Figure 4.6). The base line of each area ends east and west at the area's boundary lines.

Standard Parallels The **standard parallels (SP),** also called **correction lines,** are true parallels of latitude (east–west direction) spaced at 24-mile intervals measured from the base line. They are numbered in ascending order north or south with respect to the base line. Thus, the **first standard parallel north** is located 24 miles north of the base line (Figure 4.6). Similarly, the **second standard parallel south** is located 48 miles south of the base line.

Guide Meridians The **guide meridians (GM)** are true north–south lines. They start from the base line or a standard parallel at 24-mile intervals and run north to the next parallel of latitude that may be a standard parallel

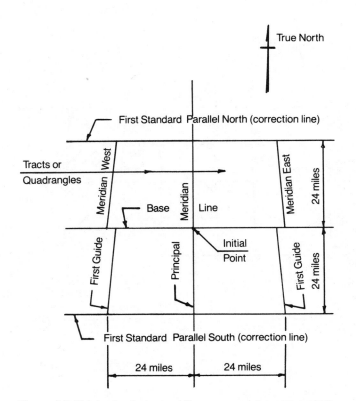

Figure 4.6 The basic elements of the government survey system.

or the base line. When they reach the following standard parallel, the distances between them become less than 24 miles because of the spherical shape of the earth. Their spacings are corrected to 24 miles as they make a new start northward to the next standard parallel (Figure 4.6). This procedure is repeated until the boundary of the area is reached. Thus, the guide meridians are not continuous straight lines. Rather, they consist of a series of lines 24 miles long. The guide meridians are numbered in ascending order east or west with respect to the principal meridian.

As noted, the corrections of the guide meridian spacings take place at the standard parallels, which is the reason for calling them correction lines.

Range Lines   Range lines, also called **meridional lines,** are true north–south meridians. They start from the base line or a standard parallel between the meridians at 6-mile intervals. They are similar to the guide meridians in that they get closer to one another as they run north until they reach the following standard parallel or base line where they get corrected (Figure 4.7).

Township Lines   Township lines, also called **latitudinal lines,** are true parallels of latitude that run between the standard parallels at 6-mile intervals. Establishing the township lines is the last step in forming the townships (Figure 4.7).

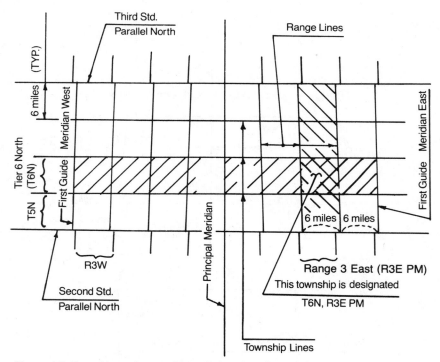

Figure 4.7 Township and range lines; designation of townships.

Township Designations   A township designation is derived from:

1. The east–west row of townships that encompasses the designated township. These rows are called **tiers** and are designated in consecutive order north or south of the *base line*. The shaded tier of Figure 4.7 is designated Tier 6 North or **T6N.**

2. The north–south row of townships that encompasses the designated township. These rows are called **ranges** and are designated in consecutive order east or west of the **principal meridian.** The shaded range of Figure 4.7 is designated "Range 3 East of the Principal Meridian" or **R3E PM.**

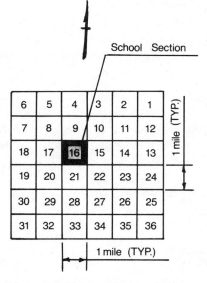

Figure 4.8 Division of a township into sections.

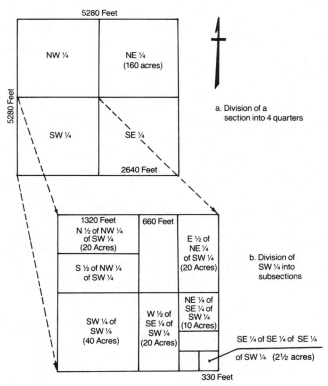

Figure 4.9 Division of a section into various subsections.

The township designation combines the designations of both of the intersecting strips. Thus, the designation of the cross shaded township of Figure 4.7 is **T6N, R3E PM.**

Sections   Each township is divided into 36 sections, each is one square mile. Their numbering sequence must be as shown in Figure 4.8. By law, section 16 of each township is set aside for school purposes, and is called the **school section.**

Subdivision of Sections   As mentioned earlier, each section is comprised of one square mile, or 640 acres. The length of each side is equal to 5280 feet. A section is divided into four quarters: North East $\frac{1}{4}$, South East $\frac{1}{4}$, South West $\frac{1}{4}$, and North West $\frac{1}{4}$, each containing 160 acres (Figure 4.9a). They are abbreviated **NE $\frac{1}{4}$, SE $\frac{1}{4}$, SW $\frac{1}{4}$,** and **NW $\frac{1}{4}$.**

Each quarter can be further divided into halves and quarters. A half of a quarter and a quarter of a quarter contain 80 and 40 acres, respectively. A quarter of a quarter can be divided into halves and quarters. For example, the NE $\frac{1}{4}$ of SE $\frac{1}{4}$ of SW $\frac{1}{4}$ contains 10 acres. Figure 4.9b shows various subsections, their identification and acreage.

Street Address

Street address may be used to identify a property in the contract of sale. However, it is not a valid method of describing land in legal documents, such as deeds and mortgages.

CHAPTER 5

Liens

A **lien** is a right or claim given by law to a creditor to the property of a debtor until the debt is satisfied. The **lienor** (the lien holder) may be entitled to sell the property to satisfy the debt with or without the consent of the owner of the property. Liens run with the land, meaning they are binding to subsequent owners if they were not discharged before transfer of title. These are the most common types of liens:

1. Mortgage lien

2. Property taxes, assessment and water charges liens

3. Mechanic's lien

4. Attachment lien

5. Judgment lien

6. Vendee's lien

7. Vendor's lien

Mortgage Lien

A mortgage is the most common type of lien. It is a **voluntary lien** in that it is created by the free will of the borrower. Almost every homeowner fi-nances a big portion of the purchase price of his house by a mortgage loan secured by a mortgage lien. The possession, control, and title of the mortgaged property remain with the debtor unless he loses these rights by a court order through foreclosure proceedings should he default on the loan.

Any interest to a real property that can be sold can be mortgaged. A homeowner can mortgage his fee simple or life estate. Similarly, a condominium owner can mortgage his fee interest in the condominium unit.

Taxes, Assessment and Water Charges Liens

These are involuntary liens created by law. They become liens to the property from the time of assessment or the time they are due until they are paid. In general, these liens take priority over all other liens regardless of the date of their recording. Taxes are levied by states, counties, towns, cities, villages, and school districts to raise revenues to perform public functions such as maintaining schools, police, fire, and sanitation departments, hospitals, etc. Property taxes are valued on *ad valorem* basis, meaning according to their value.

Assessments, or **special assessments**, are levied upon real properties in local areas to pay in full or in part for improvements that benefit these areas, such as laying sewer and water lines or paving and maintaining local streets.

Water charges are liens to the property if the water service is owned and operated by a municipality.

Enforcement of Tax Liens The most common method of collecting unpaid taxes is to sell the property. The first step is to give the delinquent taxpayer notice. If he does not respond or raise an objection, a **judgment** is rendered for the amount of taxes and interest and for the sale of the property to satisfy this amount. Subsequently, a notice of the sale is published, and the property is offered for sale at a public tax sale auction. The successful bidder is usually given a **certificate of sale** stating that he will get a **tax deed** at the end of the **redemption period** if the property is not redeemed by its owner during this period.

If the property sells for less than the taxes, interest, and sale costs, the owner is not liable for the deficit. But if the property is sold for more than the taxes, interest, and sale costs, the balance must be paid to the owner.

Tax sales must be conducted in the way prescribed in the statute. Any deviation from these procedures may invalidate the sale.

Mechanic's Lien

Few people have ever heard of a **mechanic's lien** and fewer know how it works. However, you should be aware of this lien particularly if you are planning to finance the building of your house by a construction mortgage loan. If a mechanic's lien is attached to the house during construction, it can hamper your ability to obtain subsequent loan installments.

Who Can Attach a Mechanic's Lien A mechanic's lien may be attached to your property by your laborers, material suppliers, also called materialmen, contractors or their subcontractors and material suppliers to secure payments either for work performed or material furnished where the value or condition of the property has been improved. The lien can be attached only to the property to which work was performed or materials supplied.

The right to a mechanic's lien is based on the theory that no one should be unjustly enriched. If the property has been enhanced in value due to work performed or material supplied then the workers and material suppliers must be compensated for their services. The compensation should be based on the amount of effort involved, and not on how much the value of the property has increased.

If on the other hand, the work done or material supplied did not improve the value of the property, the workers and material suppliers would not be entitled to a mechanic's lien. For example, if a delivery truck dumps a load of sand in a muddy spot, and as a result the sand could not be used, the material supplier is not entitled to a mechanic's lien.

In order for someone to attach a mechanic's lien to your property, two conditions must be met: (1) he must have a contract, either **explicit** or **implied**, with you or with your authorized representative to do the work and be paid for it, and (2) the work under contract must be virtually completed. In some states a written contract must exist; in most states it is not required.

Filing of a Notice of Lien A person who contracts directly with you can attach a mechanic's lien to your property by filing a **notice of lien** in the office of the county clerk where the property is located. The notice must be notarized under oath of the lienor or his agent.

The notice of lien may be filed at any time during the progress of work or the furnishing of the material, or within a time limit after the date of completion of the contract or the final performance of the work or the final furnishing of the material. This time limit varies between two and six months, depending upon the statute of the state. If a notice of lien is not filed within this time limit, the right to lien is forfeited.

Subcontractors and material suppliers who do not have a direct contract with you also have the right to file a mechanic's lien against your property. However, the statutes require that these persons must give you notice in writing within a specific time after their services have been rendered.

A notice of lien includes:

1. The names, residences and business addresses of the lienor(s) and the name and address of his or their attorney, if any.

2. The name of the owner of the real property, and his or her interest in the property (fee simple, etc.).

3. The name of the person by whom the lienor(s) were employed; to whom the lienor(s) furnished materials or performed professional services.

4. The kind of labor performed or the materials furnished.

5. The agreed-upon price and the value of the labor performed.

6. The agreed-upon price and the value of the material furnished.

7. The amount unpaid to the lienor(s) for labor performed and material furnished.

8. The time when the first item of work was performed; the time when the first item of material was furnished; the time when the last item of work was performed; and the time when the last item of material was furnished.

9. The location of the property subject to the lien.

If the lienor does not commence with foreclosure and sale proceedings, the mechanic's lien expires after one year from the date of filing. It may be renewed for another year by a court order.

Effect of Mechanic's Lien on Construction Loan Installments Lenders exercise due care in processing construction loan installments. Each time you apply for an installment, the lender searches the title to see whether there is a mechanic's lien attached to the property. There are two reasons for the lender's caution:

1. A mechanic's lien may be superior to all the installments paid after the filing of the lien. This may defeat the lender's intent of having a first lien mortgage to the property.

2. The intent of the law is that construction loan installments must be used to pay for the improvement of the property.

Therefore, lenders should ensure as far as possible that the laborers and material suppliers are paid.

If title search, or other actual notice, indicates that a mechanic's lien has been attached to the property, the lender will withhold installments until the lien is disposed of. The lender may satisfy (pay) the lien from the current installment and pay you the balance if you so wish.

In addition to being unable to obtain construction loan installments, you may not be able to obtain a new mortgage or sell the house as long as a mechanic's lien is attached to it.

Abuses of the Mechanic's Lien   The mechanic's lien laws are intended to protect the workers, material suppliers and contractors. However, the same laws could be abused by them because they know that in the course of building your house, you need construction loan installments and cannot wait for the lien to expire.

An example of abuse is when a contractor does some work for you without giving you a price, claiming that he cannot determine the price in advance because he does not know the extent of the work involved. After the work is done, he mails you an expensive bill far exceeding any reasonable amount. If you object or complain, he tells you that either you pay or a lien will be attached to your house. If you do not pay, he makes good on his threat and attaches a mechanic's lien to your house. All that he has to do is to file a sworn notice of lien with the office of the county clerk claiming that you have agreed to his asking price, while in fact you may not have. In many states, he is not required to present a written

contract to substantiate his claim. He can then sit back and relax knowing that he has put you in a difficult financial situation, and that you will have to submit to his terms.

Another example is when a contractor gets into a tight financial situation or goes bankrupt. You did pay him for the part of work he completed, but he did not pay his workers or material suppliers. A mechanic's lien can still be attached to your house because the law charges the owner with the responsibility of seeing that the workers are paid. Your defense to these claims will be to show proof that you have paid the contractor in full.

How to Protect Yourself against a Mechanic's Lien   You can protect yourself by having your contractors file a statement or affidavit agreeing not to file or place any liens against your property and waiving their right in this regard. If you are contracting all or a big portion of your house to a general contractor, you can include a clause in your contract with him stating that no liens on your house shall be filed by him or any of his subcontractors or material suppliers. The general contractor has to take certain legal actions, depending upon the statute of the state, such as giving timely notice of the waiver to his material suppliers or subcontractors before signing purchase orders or subcontracts. You may also withhold the final payment until the contractor provides you with an affidavit that all his subcontractors, material suppliers, and laborers have been paid in full.

If you are contracting different parts of the house to subcontractors, you may have each of them sign an affidavit stating that he will not file a mechanic's lien against your property. It is difficult, however, to have the subcontractors obtain affidavits from their workers to this effect.

The best guarantee against mechanic's liens is to investigate the reputation of each contractor before hiring him. It may be time consuming but it can save you a lot of trouble. I have found that people are usually cooperative when asked about the character or competence of a contractor they have previously hired.

Priority of Mechanic's Liens Most states treat mechanic's liens on equal footing with other liens. Their priority is determined by the order in which they are filed in the public records, with the exception of real estate taxes, special assessment, and water charges, which take priority over all other liens. Few states give preferred status to mechanic's liens, however.

Foreclosure of the Mechanic's Lien A mechanic's lien can be enforced by foreclosure proceedings whereby the lienor sues the owner and all those who recorded liens or mortgages against the property after him. If the lienor presents evidence to the court to substantiate his claims under the mechanic's lien and the court finds that you owe him a sum of money, it can order the property to be sold and the proceeds used to satisfy the debt.

If a Mechanic's Lien is Attached to Your House If a mechanic's lien is attached to your house, you have several options, depending on the situation.

1. If you owe the lienor money, you will have to pay him. Upon payment, he must give you a **certificate of satisfaction**, duly signed and acknowledged before a notary public. You must record this certificate immediately to clear the title.

2. If you do not need mortgage loan installments, a new mortgage loan, or to sell the house, do nothing until the lien expires. This is provided that the lienor does not initiate foreclosure and sale proceedings.

3. A court order to **vacate** (cancel) the lien is the course of action to take if you dispute his claim. The order may be obtained by serving notice on the lienor requiring him to either commence with foreclosure proceedings within a specific period of time (usually a month), or show cause why the lien should not be vacated (cancelled). This forces the lienor to proceed with expensive foreclosure proceedings. He must then present the court with proof of his claim against you, or he defaults and has his lien cancelled by the court.

4. Post a **surety bond** approved by the court. There are surety companies that post a bond in the amount of the lien for a fee. This frees the property from the lien. The lienor's course of action is to legally attack or foreclose on the bond.

5. Deposit an amount of money equal to the lien and interest to date of deposit in the office of the county clerk to guarantee the payment of any court judgment against you (the owner).

Attachment Lien

An attachment lien may be imposed by a person (plaintiff) upon the property of another (defendant) whom he sues for money damages. This is to prevent the defendant from selling the property that may be the only source from which to pay the damages if the plaintiff wins the case. Many states give the plaintiff this right only if the owner's residence is unknown and legal papers could not be served on him, or if he is trying to conceal the property to evade payment of judgment.

On the other side of the coin, the property owner may be unjustly hurt financially by being prevented from selling his property. Thus, statutes usually require that the plaintiff posts a surety bond to compensate the owner for any losses resulting from an attachment lien if the plaintiff does not win the case.

Judgment Lien

Money judgment is a court award for damages. A judgment lien is a lien on the debtor's property to secure the payment of the judgment. It is considered a general lien, meaning a lien to all the debtor's assets.

In many states, two conditions have to be met before a judgment becomes an enforceable lien: (1) it must be final, and (2) it must be **docketed** (recorded). A docket is a record of all judgments, and is kept in the office of the county clerk. The technicalities of how, when and where a judgment becomes a lien vary from state to state.

A judgment lien is enforced by selling the property through a **sheriff's sale**. As in other foreclosure proceedings, the sale is made public as prescribed by the statute. This is followed by a public sale auction. The purchaser

of the property must be aware that he is buying it with all the liens attached to it, with the exception of the judgment lien.

The buyer of the foreclosed property gets a **sheriff's deed** if the owner does not redeem the property.

Vendee's Lien

A **vendee's lien** may be attached by the buyer of a property (vendee) if the seller defaults. This often happens when a property is bought under land or installment contract and the seller fails to deliver the deed after he receives the agreed-upon price. The buyer is entitled to a lien in the amount he paid the seller plus any money he spent on improving the premises.

Vendor's Lien

A **vendor's lien** may be attached by the seller (vendor) to the property he conveyed to the buyer if he does not receive the price in full. The amount of the lien is equal to the unpaid balance.

Liens and foreclosure proceedings involve many technical details that vary from state to state. A lawyer's services **must** be sought in this regard. If a contractor threatens to attach a mechanic's lien to your property (if he did not provide you with an affidavit not to file a lien), you should consult a lawyer before paying him any money.

CHAPTER 6

Titles

*T*itle to real property is the right of ownership. It entitles the holder to the right of possession and control including the right to transfer ownership to others.

Origin of Titles

The origin of title depends on the geographic location. In the original 13 colonies, all title to land was from the King (or Queen) of England. To colonize this vast and scarcely inhabited land, the Crown offered huge grants to companies and individuals. Some of these individuals were: Lord Baltimore, William Penn, and the Duke of York.

In the years following the American revolution, the U.S. Government acquired about 2 billion acres (The **Public Land**) through negotiations, concessions, and purchases. The title of a portion of this public land was transferred to individuals and families (homesteaders) in order to encourage settlement and agriculture; to officers and soldiers who fought for the revolution; and to men who deserted from the British army. The U.S. government also sold land to pay for the costs of the revolutionary war.

Public Recording

Because of the significance of land ownership, the **statute of frauds** in all the states require that documents affecting the title to a property be recorded in a certain public office in the county where the property is located. Depending on the locality, the recording office may be the County Clerk, Recorder of Deeds, County Recorder's Office or the Clerk of Court of record. Almost all states require that a document be **acknowledged** before a duly authorized officer such as a **notary public** before recording. Recording has several purposes:

1. Recording gives **constructive notice** to the world of the outstanding interests in a property. Constructive notice is a fact that the buyer can discover by searching the public records. The law charges the buyer with knowing what is in the public records. Thus, he cannot claim that he did not know of a mortgage or other interest in a property if

such mortgage or interest has been recorded.

The other type of notice is the **actual notice**. It is what a buyer actually knows from inspecting the property itself or any other source. Actual notice is as good as constructive notice. If a buyer knew that the property he is buying is mortgaged but the mortgage is not recorded, his liability is the same as if the mortgage had been recorded.

2. Public recording protects the **innocent buyer** or mortgagee against previous unrecorded conveyances. An innocent buyer or mortgagee is one who paid the purchase price or loaned money without knowledge of prior unrecorded deed, or mortgage to the property.

3. Priority of liens is generally in the order of their recording with the exception of the taxes and assessments, which take precedent over all other liens regardless of their recording date. The priority of liens becomes important when the property is sold through court foreclosure proceedings for less than the amount of total liens. The tax and assessment liens are paid first, the other liens are paid in the order of their recording, leaving the lienor that recorded last with partial or no payment.

Indexing   Recording offices maintain two separate files for recorded documents. In one file, the documents are indexed in alphabetical order of the names of the grantors, mortgagors, and assignors, followed by the names of the grantees, mortgagees and assignees. In the other, the documents are indexed in alphabetical order of the names of the grantees, mortgagees, or assignees followed by the names of the grantors, mortgagors, or assignors. Thus, it is possible to locate the documents pertinent to a property if the name of a person who has an interest in it is known.

Recording Statutes

Depending upon the state, there are three different types of recording statutes:

1. **Race.** This statute gives priority to title to the first person to record the deed. If a property owner sells, or mortgages, the same property to two different buyers, or mortgagees, the first to record the deed, or mortgage, has priority to title, or lien.

2. **Notice.** This statute gives priority to title to a subsequent purchaser if he did not know at the time of purchase that there was a preceding purchaser.

3. **Race-Notice.** This statute gives priority to a subsequent purchaser if he did not know at the time of purchase that there was a preceding purchaser and if he was the first to record.

Title Search

Title search means searching the public records to discover the names of the parties who have interest in a property and to detect any defects that may affect the quality of the title. Title search reveals the entire history of a title dating back to its origin. This includes the chain of deeds, wills, mortgages, leases, and any other documents.

The average person does not have the expertise to conduct title search. Therefore, he must rely on a professional to search the title for him. Such a professional may be an abstractor, conveyancer, title insurance company, or a lawyer. The title searcher must be familiar with the records, the indexes, and the statutes applicable to situations that may arise during the search.

The search starts with the current owner in the grantee index. This leads the searcher to the name of the grantor and the book and page number in which the deed is located. Subsequently, he examines the deed, verifies the property description, and the proper execution of the deed. This constitutes the first link in the chain of title. The current grantor is the grantee in the preceding transfer of title. Next, the title searcher searches his name in the grantee index. This leads to the name of the preceding grantor and the book and page number where he can find the preceding deed. The searcher examines the documents, and this constitutes the second link in the chain of title. This procedure is repeated going back to the original grantor, and this constitutes the **chain of title** to the property. If a link in the chain is missing, it is said that there is a **gap in title**.

The title searcher concludes his work by preparing an **abstract of title**, which is a condensed history of all documents affecting the title as they appear in the public records.

Evidence of Title

Evidence of title is a proof that the seller is the owner of the property and that he has a **marketable title**. Ownership can be proven by one of the following methods: (1) an abstract of title and lawyer's opinion, (2) a certificate of title, (3) Torrens certificate of title registration, and (4) a title insurance policy. The first two methods do not guarantee against undiscovered defects in the chain of title such as forged documents, incorrect marital status or incompetent grantor. Furthermore, the statutes of limitation limit the liability of the title searchers or lawyers to a certain number of years depending on the state (about three years).

Abstract of Title and Lawyer's Opinion   Usually, the seller or his lawyer orders an up-to-date abstract of title and submits it to the buyer's lawyer who examines it thoroughly and prepares a written report on the condition of the title for his client. This report should include all the liens, mortgages, and other encumbrances up to the examination date.

Certificate of Title   A certificate of title is issued by a lawyer after he examines the public record. The lawyer writes his opinion in the title condition and lists all the existing liens and encumbrances. No abstract is prepared in conjunction with the certificate.

The accuracy of the certificate depends only on the competence of the issuing lawyer. There are no checks and balances in this method. In most cases, the issuing lawyer does not promise to defend the title against lawsuits.

Torrens Certificate of Title Registration   The Torrens certificate of title registration is a valid proof of ownership in several states. It is named after Sir Robert Torrens who introduced this system in Australia in 1858.

To register a property, its owner files an application accompanied by a substantial registration fee at the appropriate court stating his ownership to the property. After examining the application and identifying all the parties who have interest in the property, the court notifies all these parties. In addition, it publishes a notice in a local newspaper for a period of time. After careful and lengthy investigation, the court orders the title to be registered in the name of the owner. Hence, a certificate of title is issued in duplicate, one to the registrar who records it in the registration book, and the other to the owner. The Torrens certificate of title in the registrar's office reveals the name of the owner, type of ownership, and all the liens and encumbrances attached to the title, if any. However, it does not reveal certain items such as current federal and state tax liens.

To transfer title, the buyer presents the conveying document and the seller's certificate of title to the office of the registrar. After verification, the registrar cancels the seller's certificate and issues a new one to the buyer as evidence of his ownership.

The Torrens system is advantageous in that it transfers ownership at a slight expense, since there is no title search involved. However, it is not widely used because the initial registration requires lengthy legal proceedings and a hefty registration fee. Each person having interest in the property, however remote, must be notified and joined in the legal proceedings to establish ownership. The reason for this elaborate procedure is that once a title is registered, it cannot be challenged except for fraud. If a person having an interest in a property is not properly served notice, he can claim his interest after a certificate excluding him has been issued. If his claim is proven, he is compensated from an insurance fund created by the registration fees. Such action will have no effect on the title to the property.

Title Insurance Policy   A title insurance policy is issued by a title insurance company after it searches the public records and becomes satisfied that there are no apparent defects in the title. For a substantial fee to be paid at the closing, the insurance company agrees, subject to the terms of its policy, to compensate or reimburse the insured for up to the face value of the policy against any losses resulting from undiscovered defects in the title. In addition, the title insurance company agrees to defend at its own expense any lawsuit resulting from such defects. The insurance policy does not, however, cover encumbrances and defects that are discovered by searching the public records. These defects are listed in the policy as exceptions.

There are two types of title insurance policies: the **fee policy** and the **mortgage policy.** The former ensures the buyer that the property has a good and marketable title and may be conveyed by its owner. The mortgage policy ensures the mortgagee that the property has a good and marketable title and may be mortgaged by the owner (at the closing, the buyer becomes the owner).

Usually, the title insurance company issues a document entitled *Certificate and Report of Title* at the closing, and the policy at a later date. The document presented at the closing certifies that the insurance company has examined title to the premises described in an attached document called **Schedule A,** and agrees to issue its standard insurance policy after the closing. The document then specifies the type of policy (either fee or mortgage), and the amount of insurance. For a fee policy, the amount of insurance is usually equal to the purchase price of the property. However, you may want to insure the title for a larger amount since the value of the property is likely to increase with time. Naturally, more insurance costs more money. For a mortgage policy, the amount of insurance is equal to the mortgage loan.

The insurance company states that the certificate presented at the closing shall be null and void for any of the following reasons:

1. If the company's fees are not paid.

2. If the prospective insured, his attorney or agent, or the applicant or the person to whom the certificate is addressed, makes any untrue statement with respect to any material fact or suppresses or fails to disclose any material fact or if any untrue answers are given to material inquiries by or on behalf of the company.

3. In any event, upon the delivery of the policy (because the policy will supersede the certificate).

The date of the policy must be the same as the day of the closing, since the policy insures only against defects that occur before or on the date of the policy. If the certificate is dated before the closing date, the company's representative at the closing should update it to the closing date.

Usually, the title insurance company lists the following as **standard exceptions** from the coverage under its policy:

1. Any state of facts that an accurate survey might show, unless survey cover-

age is ordered. When such coverage is ordered, the certificate will set forth the specific survey exceptions that will be included in the policy. (Visual inspection of the property should reveal most if not all the defects that a land survey may indicate.)

2. Title to any personal property, whether it is attached to or used in connection with the insured premises.

3. Defects and encumbrances arising, or becoming a lien after the date of the policy, except as provided in the policy.

4. Consequences of the exercise and enforcement or attempted enforcement of any governmental, war or police powers over the premises.

5. Any laws, regulations or ordinances including but not limited to zoning, building, and environmental protection as to the use, occupancy, subdivision or improvement of the premises adopted or imposed by any governmental body.

6. Judgments against the insured or estates, interests, defects, objections, liens or encumbrances created, suffered, assumed or agreed to, by or with the privity of the insured.

The insurance company includes the following attachments as a part of the insurance certificate presented at the closing:

1. **Schedule A.** It includes the full description of the property by one or more methods so that it can be positively identified.

2. **Schedule B.** It lists all the existing liens, mortgages, covenants, conditions, easements, leases, agreements of records, etc., to the property and states that they are excepted from the policy.

3. **Tax schedule.** It shows the assessed value of the property and the city, town,

village, and school taxes and the date they are due. (This is used to divide taxes between buyer and seller at the closing.)

4. A copy of all covenants, conditions, easements, agreements of records, deed restrictions, etc., if any.

One disadvantage of the title insurance policy is its high cost, sometimes without justification. This is clearly ev-ident when an owner refinances his house and hires the title insurance company that holds the current policy to issue a new one for the new lender. The insurance company charges him the same rate as if it were searching the title going back to the origin of ownership. In reality, it searches only the period after the issue date of the current policy. During this period there are no hidden defects because there is no transfer of ownership.

CHAPTER 7

Deeds and Deed Restrictions

A deed is a document by which ownership to a property is transferred. In legal terms, a deed is a written instrument (document) duly executed and delivered, that conveys (transfers) the grantor's (seller's) right, title or interest to a property to a grantee (buyer). Duly executed means that the document is signed by the grantor and acknowledged (witnessed) by a duly authorized officer, usually a notary public, to ensure that the signature is that of the person named in the deed, and that he has signed the document with his free will.

It must be emphasized that a deed is not a proof of a marketable title, meaning a title that is free from defects that will get you, the buyer, involved in litigation (lawsuits). Rather, it is a document by which the seller transfers whatever he or she owns in the real property to the buyer.

The deed must be accurate in all its elements, otherwise it may be invalidated or challenged in court by a third party.

Essentials of a Valid Deed

A valid deed must have the following essentials:

1. The deed must be **in writing**.

2. A **grantor** (seller) that can be named with sufficient certainty to allow his identification. He must be a sane adult with the legal capacity to sign the deed. A spelling mistake in the grantor's name is not sufficient to invalidate the deed as long as the grantor can be identified. The grantor is referred to in the deed as **"party of the first part."**

3. A **grantee** (buyer) that can be named with sufficient certainty to allow his identification. As with the grantor, a spelling mistake in the grantee's name is not sufficient to invalidate the deed, as long as he can be identified. The grantee is referred to in the deed as **"party of the second part."** It is to be noted that the grantee in the current deed will be the grantor when he sells the property.

   Where there is more than one grantor or grantee, such as a husband and wife, each should be individually named in the deed. An example: "Joe Doe and Helen Doe, his wife." In case the grantor or the grantee is single, he or she is named in the deed as "Joe Doe, single," or "Helen Doe, single," so that it is documented that there is no spouse interest in the property.

4. A **consideration** (price). A deed may, depending on the state, contain a clause acknowledging the grantor's receipt of consideration. It is not necessary to state the full price of the property in the deed. It is customary to recite only a nominal consideration such as "ten dollars and other valuable consideration paid by the party of the second part." When a property is conveyed as a gift to a relative, "Love and Affection" may be sufficient consideration. Some states do not require that consideration be a part of the deed.

5. A **granting clause** (words of conveyance). This clause is extremely important since it states the grantor's intention to convey the property. The granting clause varies according to the type of deed. Typical words of conveyance are **"convey and warrant," "grant, bargain and sell," "grant and release,"** or **"grant."**

6. **Description of the property**. A deed must contain a description of the property being conveyed. The description should be adequate enough to locate the conveyed property.

7. A **habendum clause**. This is used to define the ownership to be enjoyed by the buyer. The habendum clause is **"TO HAVE AND TO HOLD** the premises herein granted unto the party of the second part, the heirs or successors or assigns of the party of the second part forever."

8. The **grantor's signature**. A deed must be signed voluntarily by the grantor(s) in order to be valid. Some states allow the deed to be signed by an *attorney-*

in-fact, meaning acting under a power of attorney, on the grantor's behalf. In this case, a power of attorney document must be recorded in the county in which the property is located. The signature must be acknowledged as evidence that it is the grantor's and that he has signed the document with his free will.

9. The **deed's delivery and acceptance**. For a title to pass, the deed has to be delivered by the grantor and accepted by the grantee.

10. **Recording**. The purpose of recording the deed is to give constructive notice to the world that the grantee (buyer) is the new owner of the property. Recording also protects the buyer if he loses the deed. Should this happen, he can obtain a certified copy of the original deed from the office of the clerk where the deed is recorded.

Types of Deeds

Deeds are not all alike. They differ according to the legal rights and guarantees contained in them. The most common types of deeds are:

1. Warranty Deed with Full Covenants.

2. Bargain and Sale Deed without Covenant against Grantor's Acts.

3. Bargain and Sale Deed with Covenant against Grantor's Acts.

4. Quitclaim Deed.

5. Tax Deed, Referee's Deed, Sheriff's Deed, and Trustee's Deed.

The Warranty Deed with Full Covenants (Warranty Deed) In this type of deed the grantor (seller) will defend the grantee (buyer) against all claims, thus providing the buyer with the highest degree of protection.

It warrants that the grantor (seller) has seized the premises in fee simple, and has the legal right to convey the same; that the grantee shall "quietly enjoy" the premises (the right to use the property without interference of possession from a third party); that the property is "free from encumbrances" (liens, easements, mortgages, etc.) except those specifically stated in the deed; that the grantor will execute any further necessary assurance to the title; and that the grantor will "forever warrant" the title to the premises.

Bargain and Sale Deed without Covenant against Grantor's Acts This deed is used to convey all right, title and interest of the seller to the buyer. The seller is not obliged to deliver a deed containing specific covenants (warrants).

Bargain and Sale Deed with Covenant against Grantor's Acts This is the type most often used to transfer ownership of real property. It is identical to the bargain and sale deed without covenant against grantor's acts in every respect except that the grantor promises that he has done nothing to encumber the property while it was in his possession. In legal terms: "And

the party of the first part [grantor] covenants that he has not done or suffered anything whereby the said premises have been encumbered in any way whatever."

Quitclaim Deed The usual purpose of this form of deed is to remove a **cloud** (outstanding claim or encumbrance that, if valid, would affect or impair the title to a property) from the title to a property whereby a person (or persons) who may have some claim to the property is asked to release his interest (**"quitclaim"**). Its form is identical to that of the bargain and sale deed without covenant against the grantor's acts except that it uses the words **"remise, release** and **quitclaim"** instead of the conveying words "grant and release."

Tax Deed, Referee's Deed, Sheriff's Deed, and Trustee's Deed These deeds are used to convey title to properties that are sold pursuant to court orders in foreclosure proceedings to satisfy unpaid taxes, mortgages, and judgments.

Deeds are executed on printed forms. The type of deed is clearly indicated by fancy labelling printed on the back of the form. Figure 7.1 shows the la-

Warranty Deed
WITH FULL COVENANTS

Bargain and Sale Deed
WITH COVENANT AGAINST GRANTOR'S ACTS

Quitclaim Deed

Figure 7.1 Different types of deeds.

belling of several types of deeds. More information on the tax deed can be found in Chapter 5; on the referee's and sheriff's deed in Chapter 8; and on the trustee's deed in Chapter 9.

Deed Restrictions

It should be clear that the deed is pertinent to the land (lot or parcel) and its use. This includes the buildings that are on it or to be constructed on it. Any land owner can impose deed restrictions by including them in the deed when the property is conveyed (sold). Deed restrictions are beyond and above zoning ordinances set forth by the local governments. An example of deed restrictions is when a land subdivider sets restrictions regarding land use such as the type of building, type of exterior siding, minimum square footage, height of buildings, setbacks, etc. For deed restrictions to be valid, they have to be reasonable and to the benefit of all lot owners in the subdivision. Each lot owner has the right to obtain a **court injunction** to prevent the owner of a neighboring lot from violating the deed restrictions. Generally, deed restrictions have a time limit beyond which the majority of the owners in the subdivision may change or modify the terms of the restrictions.

Deed restrictions may be included through a covenant in the deed if they are few, or in a separately executed and recorded declaration if they are many. A grantor should bear in mind that the more restrictions he imposes, the less desirable the property becomes.

Deed restrictions can be elaborate and involve minute details. For example, the deed restrictions of my own lot are contained in a separate declaration comprising four long sheets. Excerpts of this declaration are shown below to demonstrate how elaborate deed restrictions can be.

No building or structure other than a private dwelling, not to exceed 2½ stories, designed for occupancy and use for one family only, shall be erected, maintained or permitted . . . ; no portion of any residence nor of any piazza stoop or porte cochere thereto, shall be erected within 40 feet of the front and rear lines, or within 20 feet of either of the side lines of said premises; and that no clothes line, or other clothes drying equipment shall be erected or maintained upon said premises unless the same be so screened or enclosed as to keep the same and all clothes drying from the view of the occupants of all neighborhood property; and all ashes, refuse and garbage shall be kept in receptacles which are all metal, water tight and closed and such receptacle shall, at all times, be screened from view and the contents thereof shall not be scattered over any part of the premises and shall not be disposed of on the premises, except by internal incineration, and shall not be buried on the premises; that sewer, gas, water and other house and street connections, excepting electric and telephone connections, shall be placed underground . . . ; nor shall any animals, which may be offensive or create a nuisance, including poultry, chickens, or the like, or any dangerous dog be kept thereon nor shall any sign or signs be placed on any part of the residence other than the usual "for sale signs" without the consent in writing of the undersigned corporation or its successors, or assigns.

That these covenants are to run with the land and shall be binding on all parties . . . until January 1, 1981, at which time said covenants shall be automatically extended for successive periods of ten years unless by a vote of a majority of the then owners of the lots or plots affected by these restrictions, it is agreed to amend, modify or change the said covenants in whole or in part. . . .

That if the owners affected by these covenants and restrictions or any of them or their heirs or assigns shall violate or attempt to violate any of the covenants herein it shall be lawful for any other person or persons owning any real property situated in said development or subdivision to prosecute any proceedings at law or in equity against the person or persons violating or attempting to violate any such covenants and either to prevent him or them from so doing or to recover damages or other dues for such violation.

CHAPTER 8

Mortgage Law

Mortgages are long-term loans used to finance real property purchases, whereby the borrower pledges his ownership in the property as collateral to secure the debt. The rights of the lender to the mortgaged property depend on the mortgage theory of the state in which property is located. There are three theories, each adopted by a number of states:

1. **Lien theory states.** These states consider the mortgage to be a lien to the property. Accordingly, if the borrower defaults, the lender has to obtain a court order to sell the property through foreclosure proceedings to collect his debt. There are 32 lien theory states.

2. **Title theory states.** These states consider that a mortgage transfers some title rights from the borrower to the lender. According to this theory, the lender has the right to possess and rent the property. However, he does not have the right to ownership. To acquire this right, the lender must go through court foreclosure proceedings. The title theory is followed by 14 states and the District of Columbia.

3. **Intermediate theory states.** These states take a middle course between the lien and title theory states. In the intermediate theory, the lender has the right to possess and rent the property only if the borrower defaults. There are four states in this category.

A mortgage loan is comprised of two documents: the **promissory note** or **note** and the **mortgage**. The promissory note includes the debt and the mortgage creates the lien that secures the debt. Both documents are necessary to create an enforceable mortgage loan. In some states, a **bond** is used instead of a note to accomplish the same objectives.

The Promissory Note

The promissory note is the evidence of debt. It usually includes the amount of the debt (loan), the interest rate, method of payment and the borrower's promise to repay the loan plus interest to the lender.

Essentials of a Promissory Note A valid note must be in writing and must include:

1. A borrower and a lender, both with contractual capacity.

2. A promise by the borrower to pay a specific sum of money.

3. The terms of payment.

4. A clause stating that the note is secured by a mortgage (without this clause, the loan would be a personal loan).

5. The borrower's signature.

The note must be voluntarily delivered by the borrower and accepted by the lender.

The interest is not one of the essentials; however, if interest is charged, it must be included in the note.

Printed Promissory Note Form Most lenders use a printed form conforming to the standards of the **Federal National Mortgage Association/Federal Home Loan Mortgage Corporation (FNMA/FHLMC)** to execute the promissory note. This would be required if the lender were to sell the note to any federal or quasifederal agency. The form includes:

1. The location and date the note was signed.

2. The borrower's promise to pay, in return for the loan received, a specific sum of money called the principal, plus interest to the lender. The principal is usually equal to the face value of the loan. The lender is then identified. Next, a clause that the borrower understands that the lender may transfer the note (sell it in the secondary mortgage market). The lender or anyone to whom the note is transferred is called the **note holder.**

3. The borrower's agreement to pay the interest rate specified in the note. Interest is to be paid on the outstanding part of the principal only.

4. The borrower's agreement to make monthly payments of the principal and interest, in the amount specified in the note, until the loan is amortized. The monthly payment is due on the first day of each month beginning with the second month after the closing. (Some states give the borrower a grace period of a certain number of days after the first day of the month during which he can pay without penalty.)

5. The borrower's agreement to pay the stated late charge for overdue payments. (Usually 4 or 5 percent of the overdue amount of principal and interest.)

6. The borrower's agreement that if he does not pay the full amount of each monthly payment on time, the lender may send him a written notice telling him that unless he pays the overdue amount by a certain date, not less than 30 days from the date of mailing the notice, he will be **in default.**

7. The borrower's agreement that if he does not pay the overdue amount by the date stated in item No. 6, he will be in default. The note holder may require the borrower to pay immediately the full amount of outstanding principal and all the interest (this is called the **acceleration clause**).

8. The borrower's agreement that if he defaults, the lender will have the right to be paid back for all reasonable costs and expenses including reasonable attorney's fees.

9. A clause stating that the borrower has the right to make prepayment of the principal in full or in part. (Some lenders impose penalties for prepayments if they were made within a specific period after the closing date, subject to the limitations of the federal and state governments. You should ask about the lender's policy regarding prepayment penalties before you apply for a mortgage.)

10. A clause stating the mailing address of the borrower to which any notice will be delivered.

11. A clause stating that this note is covered by a mortgage that protects the lender from possible losses that might result if the borrower does not keep the promises he made in this note.

The note must be signed by the borrower. However, it may or may not be required that the signature be acknowledged before a notary public depending on the laws of the state (usually not). The lender does not sign the note.

Some lenders may include other conditions than those of the **FNMA/FHLMC** printed form, in a separate document called rider to the note. Such a rider becomes null and void if the lender sells the note to a federal or quasifederal agency.

The Mortgage Document

The mortgage is a document whereby the borrower pledges a property as collateral that the lender can sell to secure the payment of the amount in the note as agreed upon. Most lenders use a FNMA/FHLMC printed form to execute the mortgage for the same reasons explained in the section on promissory notes. In mortgage agreements, the borrower is called the **mortgagor** because he, or it, is the one who mortgages his property; and the lender is called the **mortgagee** because he, or it, is the one who receives the mortgage. The mortgage form is more elaborate than that of the note. It includes three main parts: (1) mortgage, (2) uniform promises, and (3) non-uniform promises.

Mortgage   This part is divided into four sections: I. Words used often in this document; II. Borrower's transfer of rights in the property to lender; III. Description of the property; IV. Borrower's right to mortgage the property and borrower's obligation to defend ownership of the property.

I. Words Used Often in this Document.
   A. The date of signing the document (the closing date).
   B. The name of the borrower(s). If the borrowers are a husband and wife, it should be so stated (such as Joe Jones and Helen Jones, his wife). The address of the borrower is also indicated.
   C. The name and address of the lender.
   D. A reference to the note signed by the borrower, the amount of debt shown in the note, and the borrower's promise to pay that amount in full plus interest by the end of the term of the loan.
   E. A clause stating that the property described in the section entitled "Description of the Property" will be called "the property."
II. Borrower's Transfer of Rights in the Property to Lender.
   (This section is the core of the mortgage because it includes the borrower's pledge of his property to the lender.)
   I mortgage, grant, and convey the property to lender subject to this mortgage.

This means that, by signing this mortgage, I am giving lender those rights that are stated in this mortgage and also those rights that the law gives the lenders who hold mortgages on real property. I am giving lender these rights to protect lender from possible losses that might result if I fail to:

A. Pay all the amount that I owe lender as stated in the note;

B. Pay, with interest, any amounts that lender spends under this mortgage to protect the value of the property and lender's rights in the property;

C. Pay, with interest, any other amounts that lender lends to me as future advances under paragraph 23 titled "Agreements About Future Advances"; and

D. Keep all of my other promises and agreements under this mortgage.

It is to be noted that the words of conveyance conform to the **title theory** position. Nonetheless, in **lien theory states,** these words give the lender only a lien to the property. This is evidenced in the words "Those rights that the law gives to lenders who hold mortgages."

III. Description of the Property.

The location or address of the property is given in this section. The county or city where the property is located is identified. The property's **legal description** is given in detail in a separate document called "Schedule A".

IV. Borrower's Right to Mortgage the Property And Borrower's Obligation to Defend Ownership of the Property.

A. The borrower promises that aside from the "exceptions" listed in the title insurance mortgage policy (1) he lawfully owns the property, (2) he has the right to mortgage, grant, and convey the property to lender, and (3) there are no outstanding claims or charges against the property.

B. The borrower gives a general warranty of title to lender. This means that the borrower will be fully re-

sponsible for any losses that the lender suffers because someone else has some of the rights to the property that the borrower claims he has. The borrower promises that he will defend his ownership of the property against any claims of such rights.

Uniform Promises The uniform promises are used in mortgages all over the country. These are the highlights:

1. **The borrower's promise to pay.** The borrower shall promptly pay when due: principal, interest, and late charges as stated in the note. Also principal and interest on future advances that he may receive under paragraph 23.

2. **Escrow account (the funds).** The borrower agrees to pay to lender all amounts necessary to pay taxes, assessments, and hazard insurance on the property, unless the lender tells him in writing that he does not have to do so, or unless the law requires otherwise. This money will be called the "funds." The lender is obliged to keep the funds in an **escrow account.**

The lender is not required to pay the borrower any interest or earnings on the funds unless both agree at the time of signing the mortgage that the lender will pay interest on the funds; or if the law requires lenders to pay interest on the funds. (Some states require that interest be paid on the funds in the amount of 1 or 2 percent, annually.)

If lender's estimate of funds is too high, the borrower has the right to have the excess amount either refunded to him or credited to his future monthly payment of funds. (The maximum amount of funds that a lender can withhold in the escrow account is regulated by the **Real Estate Settlement Procedures Act or RESPA** as explained in Chapter 13.)

3. **Application of payments.** Unless the law requires otherwise, lender will apply each of the payments in the following order: (1) escrow account for taxes and insurance, (2) interest, (3) principal, and (4) interest and principal on any future advances.

4. The borrower will pay all taxes, assessments and any other charges and fines; and will satisfy all liens against the property that may be superior to this mortgage.

5. The borrower is obliged to obtain and keep a hazard insurance policy on the property.

6. The borrower is obliged to keep the property in good repair, not to destroy, damage or substantially change the property and will not allow the property to deteriorate.

7. The lender has the right to protect the property against legal actions by others. The borrower must pay the lender all the money it spends plus interest.

8. The lender has the right to inspect the property. Before an inspection is made, the lender must give the borrower notice stating a reasonable purpose for the inspection.

9. The lender has the right to the proceeds from all awards if the property is taken by any governmental authority through **condemnation.**

10–16. Clauses addressing continuation of the obligations of the lender, borrower, persons taking over borrower's rights or obligations; and that the borrower shall be given a

copy of the note and of the mortgage either when the note and mortgage are signed or after the mortgage has been recorded in the proper official records.

17. **VA-guaranteed loan.** If the loan is a VA-guaranteed loan, the rights and obligations of both borrower and lender are governed by the law known as Title 38 of the United States Code and Regulations, also called the **VA Requirements.** If there is any conflict between one or more terms of the mortgage and the VA requirements, the mortgage terms shall be modified or eliminated to agree with the regulations.

18. If the lender requires a mortgage insurance as a condition of making the loan, the borrower shall pay the premium of such policy until the requirement for mortgage insurance ends according to written agreement with lender or according to law.

Non-Uniform Promises The non-uniform promises vary to a limited extent in different parts of the country. The highlights of these promises are:

19a. **Assumption of mortgage.** If the borrower transfers all or part of the property or any rights in the property, the person to whom the property is sold may take over the borrower's right and obligations under the mortgage (this is known as **assumption of the mortgage**), if these conditions are met: (1) the borrower gives the lender notice of the sale or transfer, (2) the lender agrees that the person's credit is satisfactory, (3) the person agrees to pay interest on the amount owed at whatever rate lender requires, and (4) the person signs an assumption agreement that is acceptable to lender. If all these conditions are met, the lender will release the borrower from all obligations under the note and mortgage.

19b. Lender's rights if borrower transfers the property without meeting conditions of paragraph 19a. If the borrower sells or transfers the property and conditions (1), (2), (3) and (4) of paragraph 19a are not satisfied, the lender may mail the borrower a notice requiring the immediate payment in full of the entire amount under the note. The notice will give the borrower at least 30 days to make the required payment. If the borrower does not make the required payment during that period, lender may bring a lawsuit for **foreclosure and sale** without giving the borrower any further notice or demand for payment.

20. Lender's rights if borrower fails to keep promises and agreement. If the borrower fails to keep any promise or agreement made in the mortgage, and if the lender sends the borrower a notice that states the failures and the actions needed to correct them and if the borrower does not correct the failures stated in the lender's notice, the lender may require the immediate payment of the principal and interest in full and may bring a lawsuit to take away all the borrower's remaining rights to the property, and to have the property sold through foreclosure and sale proceedings. The lender also has the right to collect all costs of the lawsuit as allowed by law.

21. Borrower's right to have lender's lawsuit for foreclosure and sale discontinued. Even if the lender has required immediate payment of the loan in full, the borrower has the right to have any foreclosure and sale lawsuit or other enforcement of the mortgage discontinued any time *before a judgment has been entered* enforcing the mortgage if he meets the following conditions:

a. Pays to lender the full amount that would have been due under the mortgage, the note and any notes for future advances.

b. Corrects his failure to keep any of his other promises or agreement made in the mortgage.

c. Pays all of the lender's reasonable expenses in enforcing the mortgage.

d. Does whatever the lender reasonably requires to ensure that the lender's rights in the property and his obligations under the note and the mortgage will continue unchanged.

If the borrower fulfills all four items mentioned above, then the note and the mortgage will remain in full effect *as if immediate payment in full had never been required.*

22. The lender has the right to rental payments from the property and to take possession of the property. (Again, this is the language of the title theory states. However, the theory applied in the state where the property is located shall prevail.)

23. Agreements about future advances. The borrower may ask the lender to make one or more loans to him in addition to the loan included in the note. The lender may, before the mortgage is discharged (cancelled) make those additional loans to the borrower. The mortgage will protect the lender from possible losses that might result from the borrower's failure to pay the amounts of any of those additional loans plus interest, only if the notes that contain the borrower's promises to pay those additional loans state that the mortgage will give lender such protection. Additional loans made by the lender that are protected by the mortgage will be called **"future advances."** (The maximum amount of the future advances, if any, should be stated in this paragraph.)

24. **Satisfaction of mortgage.** The lender is obliged to discharge (cancel) the mortgage when he is paid all amounts due under the note, the mortgage and any notes of future advances. The lender will discharge the mortgage by delivering a **satisfaction of mortgage certificate** stating that the mortgage has been satisfied. The borrower is not required to pay the lender for discharge, but he shall pay all costs of recording the discharge in the proper official records. (The satisfaction of mortgage certificate should be recorded without delay to clear the title.)

The mortgage document must be signed by the borrower(s) and acknowledged before a notary public since it will be recorded. As in the case of the note, the lender does not sign the mortgage document.

Default and Foreclosure

Most defaults occur when a borrower fails to make payments on his mortgage loan as stated in the promissory note and the mortgage document. When this happens, the loan is said to be **delinquent.** Defaults can also result from failure to pay property taxes,

assessments, or through the breach of any of the conditions contained in the mortgage document. Should the loan become delinquent, the lender has several legal options available to it to protect its money. However, most lenders prefer to meet with the borrower to arrange for a new overstretched payment plan before starting lengthy and costly legal actions. If the borrower realizes that he cannot meet the proposed plan, he may be pursued to sell the property rather than have it foreclosed upon. It is when the borrower cannot find a buyer who is able and willing to pay a price equal to the debt, that the lender invokes the acceleration clause. Subsequently, the lender may proceed with one of the legal remedies that the laws of the state allow. The most common remedies are: (1) legal foreclosure, (2) power of sale, (3) deed in lieu of foreclosure, and (4) strict foreclosure.

Legal Foreclosure and Sale   The legal foreclosure and sale proceedings begin with a title search to identify all the persons who have interest (the lienors, mortgagees, etc.) in the property. Next, the lender files a lawsuit naming the borrower and all those who recorded their interest in the property subsequent to the foreclosed upon mortgage as defendants. Lienors or mortgagees who recorded their interest before the mortgage that is being foreclosed upon have senior claims and are not affected by the lawsuit.

The lender's complaint describes the note or bond, the mortgage that secures it, the amount of debt (principal and interest), and that the loan is in default. Next, the lender asks the court for a judgment directing *the cut-off of the interest of all the defendants,* the sale of the property, and the application of the proceeds of the sale to pay the debt and expenses.

Lis pendens In the meantime, a notice of the lawsuit is filed in the office of the county clerk where the property is located. This notice, called lis pendens, is to warn the public that a legal action is pending against the property. The borrower can still sell the property after a lis pendens has been filed. However, prospective buyers, who would undoubtedly search the public records, would learn about the pending lawsuit and price the property accordingly.

Next, a copy of the summons and complaint is served on the borrower and the other defendants, if any. They are given a certain period of time during which they may reply by presenting their side of the argument to the court judge. If neither the borrower nor any of the other defendants replies, or if their reply is found to be in favor of the lender, the judge will order the property to be sold, through a public auction.

The sale is made public by posting a notice on the courthouse door and advertising it in local newspapers, as prescribed by the law. In addition, all the defendants are given notice of the sale. The sale is held at either the property or the steps of the courthouse, and is conducted by a court-appointed **master, county sheriff** or **referee.**

Equitable right of redemption All states give the borrower or any other defendant the right to redeem his property until the sale starts if he pays the principal, interest, and costs of sale. This right of the owner (or any junior lienor) to redeem the property is called **equitable right of redemption** or **equity of redemption.** This right is cut off when the sale takes place.

The sale. If no one redeems the property, the sale begins. Any person can bid provided that he can pay the price in cash. The highest bidder (the buyer) is required to pay a cash deposit equal to 10 percent of the price at the auction, and the balance at the closing (in about 30 days). The sale must be approved by the court, which has the power to order a new sale if it feels that the bidding price is too low.

In the states that do not have statutory redemption (explained below), the highest bidder receives a **referee's deed** or a **sheriff's deed** when he pays the full price. In this deed, the lien that is the cause of the sale and the court order to sell the property are stated, and the referee or sheriff conducting the sale is the grantor.

The sale expenses are the first item to be paid from the sale proceeds. Next, the liens are paid according to the dates of their recording. If the sale price is more than the expenses and debt, the remaining balance is paid to the borrower.

Deficiency judgment If the sale price is less than the debt and expenses, the lender may obtain, depending on the laws of the state, a deficiency judgment against the borrower. This judgment gives the lender the right to collect the debt against the borrower's real and personal assets.

Statutory redemption Many states have statutory redemption laws that allow the borrower a statutory redemption period after the date of sale, during which he can redeem the property. This period varies between two months to two years, depending on the state. Furthermore, some states give the borrower the right to possess the property during the statutory redemption period. The highest bidder in the sale receives a **certificate of sale** that entitles him to a referee's deed or sheriff's deed if the borrower does not redeem the property.

Power of Sale   The power of sale, also known as **foreclosure by advertisement,** or **sale by advertisement,** is not a court proceeding. Rather, the statute in several states gives the lender the right to sell the mortgaged property after giving a notice to the borrower. Hence, the property is advertised for sale and then sold. Generally, the lender who resorts to power of sale loses the right to a deficiency judgment, should the property sell for less than the debt.

The statute of each state sets specific procedures that must be followed when foreclosing on a property through the power of sale. One of these procedures is to allow for a waiting period between default and the sale. This is to give the

borrower time to repay the debt and redeem the property.

Deed in Lieu of Foreclosure   If the borrower is convinced that his property will sell for less than the debt, and if the lender is convinced that the borrower has no other assets, personal or otherwise, that warrant a deficiency judgment, they both may agree that the borrower delivers and the lender accepts the deed to the property.

The lender must be prepared to prove that it gave the borrower a fair deal and did not take advantage of his financial difficulty. By accepting the deed, the lender assumes the responsibility to pay the outstanding taxes and all other lienors. If the property is valued higher than the debt, the lender must pay the borrower the difference in cash.

Strict Foreclosure   Strict foreclosure is applied in a few of the title theory states. There is no public sale involved. The lender asks for a strict foreclosure, and if the borrower does not object, he loses the property to the lender. If he does object, the court orders a sale.

CHAPTER 9

Types of Mortgages

There are several types of mortgages, each best suited for a specific purpose or situation. The type of mortgage or mortgages you select to finance your new house depends on several factors. Some of these factors are personal; for instance, if you are a veteran you may qualify for a VA-guaranteed mortgage loan. Another factor is how the house is going to be built. If you contract a developer to build a house including the lot, you can finance the deal by one long-term mortgage loan. But if you buy a lot and hire either a general contractor or several subcontractors to build the house you need two mortgage loans: (1) a building or construction loan, which is a short-term or interim loan to finance construction until the house is built and approved for occupancy, and (2) a long-term loan to refinance the building loan. You may have both loans approved in one closing. These are the most widely used types of mortgage loans:

1. Conventional fixed-rate mortgage
2. Conventional adjustable-rate mortgage
3. FHA/HUD-insured mortgage
4. VA (GI)-guaranteed mortgage
5. Building or construction loan mortgage
6. Deed of trust mortgage
7. Blanket mortgage
8. Reverse mortgage
9. Open-end mortgage
10. Purchase money mortgage
11. Package mortgage
12. Graduated payment mortgage

Conventional Fixed-Rate Mortgage

The term **conventional mortgage** may be defined as a mortgage loan that is made according to the lender's own regulations. It may also be defined as a loan that is not FHA-insured or VA-guaranteed.

In **fixed-rate mortgages,** the interest rate is established when the loan is made and does not change throughout the life of the loan. It used to be the only way of charging interest, until 1980 when interest rates went up sharply and lenders suffered heavy losses because they had to pay up to 16 percent interest rates to raise funds while their old fixed-rate mortgages were paying only about 8 percent. To reduce the lender's risks, the government authorized them to make loans with variable or adjustable-interest rates. Nevertheless, lenders continue to offer fixed-rate mortgages because many borrowers do not feel comfortable with adjustable-rate mortgages. Understandably, they charge higher rates or more points on fixed-rate than on adjustable-rate mortgages to protect themselves against a possible rise in interest rates. Naturally, if interest rates go down, fixed-rate lenders make more profits.

Conventional Adjustable-Rate Mortgage

The **adjustable-rate mortgage,** known as **ARM,** is a loan whereby the interest rate changes periodically based on a certain monetary index. In most ARM loans the monthly payments change following the rise and fall of interest rates. Adjustable-rate mortgages have several variables other than interest rates. You need to know what these variables are in order to be able to compare one ARM with another or compare adjustable with fixed-rate mortgages. The most important variables are the adjustment period, index, margin, interest cap, payment cap, negative amortization and conversion.

The Adjustment Period The adjustment period varies widely, depending on the lender and on the volatility or stability of interest rates at the time of making the loan. Some loans call for an adjustment every month, others every few years. It is very hard to tell which is in the borrower's best interest because no one can predict the course of interest rates during the following 15 to 30 years. As a general rule, if you think that the rate at the time of making the loan is low, you may choose a loan with long adjustment period to lock in the low rate for several years. Conversely, if you feel that interest rates may come down, you may choose a loan with short adjustment period.

The Index Each lender ties its ARM interest rate to the rate of a particular **index.** Such an index may be the prime interest rate; the national or regional average cost of funds to savings and loan associations; the interest rate on the latest bid on two- or five-year treasury notes or bonds; the latest bid on three- or six-month treasury bills; or the lender's own cost of funds.

The Margin After establishing the index, each lender adds a few percentage points to the rate of the index to cover its profit and overhead expenses. The added percentage is called the **margin** and it varies from one lender to another. Thus, the ARM interest rate is equal to the rate of the index plus the margin.

Interest Cap The interest cap in ARM means a limit on the increase in the interest rate per adjustment period. There are two types of caps: periodic and overall. **Periodic cap** limits the increase in interest rate for each adjustment period. **Overall cap** limits the interest rate increase throughout the life of the loan. As an illustration, let us assume that an ARM has an initial interest rate of 10 percent, adjustment period of one year, periodic cap of 2 percent and overall cap of 5 percent. The maximum interest rate that could be charged on such a loan after one year is 12 percent; after 2 years is 14 percent; after 3 years and beyond is 15 percent. If the index moves down, the interest rate on the loan moves down accordingly.

Having an ARM with a cap protects the borrower against drastic increases in monthly payments should interest rates increase sharply. Mortgage loans are long-term with life spans varying between 15 and 30 years. During such a period, the movement of interest rate cannot be predicted.

Payment Cap Some lenders offer adjustable-rate mortgage loans with payment caps that limit the increase in the monthly payment for each adjustment period regardless of the increase in the interest rate charged on the loan. This method does not relieve the effect of a significant increase in interest rates. It just postpones and compounds the problems because it causes negative amortization of the loan.

Negative Amortization Negative amortization means that the balance of the principal increases rather than decreases each time a payment is made. This happens to ARM loans with payment caps when the allowable increase in the payment does not catch up with the increase in the interest rate to the extent that the payment does not satisfy all the interest. The deficit is then added to the principal causing it to increase. As a result, the interest on the subsequent payment will be higher causing more deficit and higher principal, and so on.

Conversion Some ARM loans have a clause by which the variable-rate can be converted to a fixed-rate loan at a designated time. The lender should state at the outset how the fixed-rate will be determined, and whether there is an additional charge for exercising this option.

More information about adjustable-rate mortgages is included in a booklet prepared by the Federal Reserve Board and the Federal Home Loan Bank Board entitled *Consumer Handbook on Adjustable Rate Mortgages*. Lenders are required to mail a copy of this booklet with the application form to their prospective borrowers.

FHA/HUD-Insured Mortgage

The Federal Housing Administration (FHA) is a branch of the **U.S. Department of Housing and Urban Development (HUD).** It was created by Congress in 1934 to stimulate new construction after the depression. It began by offering to insure lenders against losses due to the borrower's default if lenders would agree to lend up to 80 percent of the equity value of new or existing houses for up to 20-year amortized loans. (In those days, the average loan-to-value ratio was about 40 percent and the loan's term was 5 years.) In addition, the FHA contributed to improving the quality of construction by setting minimum standards and sending its inspectors out to determine that the houses met those minimum standards before it agreed to insure the loans. To cover its possible losses, the FHA charged the borrower an annual insurance fee, based on the outstanding balance of the loan.

Nowadays, FHA/HUD has a diverse mortgage insurance program. Under this program the department does not provide direct government mortgage financing. Rather, it insures mortgage loans made by local lenders. There is a top limit for the amount of insured loan depending upon the location of the property and the number of units in the dwelling. If the price of the property is higher than the top limits set by the FHA, the buyer has to pay the difference in cash.

To obtain one of the FHA-insured loans, one applies for it at one of the FHA/HUD–approved lenders. The lender then processes the application according to FHA procedures. The property must be appraised by an FHA-approved appraiser. The applicant must have satisfactory credit and sufficient income to support the mortgage and pay for other living expenses and all outstanding obligations. FHA-insured loans may be assumed by the new buyers without obtaining FHA approval.

Mortgage Insurance Premium The FHA charges the borrower an insurance premium that depends on the type of the loan. For loans under Section 235 (explained in the following), the annual premium is 0.7 percent of the outstanding balance. For loans under other sections, the premium is a one-time charge of approximately 3.8 and 2.9 percent of the value of the loan for 30- and 15-year loans, respectively, payable at the closing. A part of this premium is refunded if the loan is paid within the first 10 years of the life of the loan.

The most important sections under the FHA/HUD program are:

Section 203 B This is the most popular type of the FHA/HUD–insured loans. It insures one- to four-family dwellings. In the least expensive counties, the current top limits of the insured loans are $67,500, $76,000, $92,000 and $107,000 for one-, two-, three-, and four-family dwellings, respectively. In the most expensive counties, the current top limits are $90,000, $101,300, $122,650 and $142,650 for one- to four-family dwellings, respectively.

The maximum amount of the insured mortgage loan under this section is 97 percent of the first $25,000 of the appraised value of the property, plus the closing costs, plus 95 percent for the portion of the price in excess of $25,000. If the loan is for refinancing, the maximum amount of the insured mortgage is equal to 85 percent of the appraised value of the property plus the closing costs.

Section 221 D2 The loans insured under this section are for single family homes only. The borrower (mortgagor) must occupy the premises and must qualify as a family. The maximum insured amount is 97 percent of the appraised value of the property plus the closing costs. The maximum insured mortgage is $36,000, which can be increased to $42,000 for four-bedroom homes with five or more people in the family.

Section 222 This section provides insured mortgage loans for single family homes for active-duty servicemen. The **Department of Defense** pays the mortgage insurance premium. The maximum insured amount is the same as under Section 203 B.

Section 245 The loans insured under this section are **graduated payment**

mortgages (see page 67). They are for single family homes only. The maximum insured amount is the same as under Section 203 B. The mortgage payments are less than the interest for a major portion of the graduating years. The portion of interest that is not paid during the graduating period is added to the principal (negative amortization). There are five authorized payment plans under this program:

Plan I: Monthly mortgage payments increase 2½ percent each year for five years.

Plan II: Monthly mortgage payments increase 5 percent each year for five years.

Plan III: Monthly mortgage payments increase 7½ percent each year for five years.

Plan IV: Monthly mortgage payments increase 2 percent each year for ten years.

Plan V: Monthly mortgage payments increase 3 percent each year for ten years.

Section 235 This program is called "Home Ownership Assistance Program." The application for mortgage insurance under this program must be submitted to the FHA field office. The borrower must find a lender who is willing to (1) make the loan, (2) receive and maintain records of monthly assistance payments on behalf of eligible homeowners, (3) obtain the required periodic recertification of family incomes every year, and (4) make appropriate adjustments in billing for monthly assistance payments. The lender is compensated for the additional work involved as authorized by law.

The assistance is in the form of monthly payments by the Department of Housing (HUD) to the lender. Family incomes and the maximum insured

limits are established for each locality. The maximum mortgage amount is $40,000 and $47,500 for low and high priced areas, respectively.

A single family dwelling must be new or substantially rehabilitated and approved for FHA mortgage insurance prior to the beginning of construction or substantial rehabilitation.

VA (GI)-Guaranteed Mortgage

In 1944, the Congress passed the **Servicemen's Readjustment Act** whereby the **Veterans Administration (VA)** was authorized to guarantee loans used to buy or build homes for eligible veterans of World War II. This way, the veterans could buy homes with no down payment.

At present, the VA guarantees a portion of the mortgage loan in the amount of $27,500 or 60 percent of the value of the loan, whichever is less. The $27,500 guarantee allows the veteran to obtain a mortgage loan of up to $110,000 with no down payment, provided that his income can support the monthly payment (the guaranteed $27,500 represents 25 percent down payment for a $110,000 loan). A veteran may buy a house for a higher value than $110,000 provided that he pays the difference in cash. Loan applications must be submitted at a VA-approved lender. If there is no nearby available lender, the VA provides the loan.

The rules for VA-guaranteed loans change from time to time. A veteran who wants to know what guarantees he or she is entitled to should apply at the VA for a **certificate of eligibility**, which is one of the documents required to obtain a loan. The VA will mail interested veterans several pamphlets explaining its loan-guaranteed plans as well as tips on what to look for when buying a house. Some of these pamphlets are: (1) *To The Home-Buying Veteran*, a guide for veterans planning to buy or build homes with a VA loan, (2) *Pointers For The Veteran Homeowner*, a guide for veterans whose home mortgage is guaranteed or insured under the GI bill, and (3) *VA-Guaranteed Home Loans For Veterans*.

To protect the veteran and minimize the possibility of his default, the VA assigns an approved appraiser to value the property before the loan is made. Based on his appraisal, the VA issues a **Certificate of Reasonable Value (CRV)** informing the veteran of the appraised value of the property, and the maximum VA-guaranteed loan that a lender can give him. If the price of the house is greater than the appraised value shown in the CRV, the veteran can still buy it, but he will have to pay the difference in cash.

The VA-guaranteed loan may be for up to a 30-year term. The maximum interest rates are set periodically by the VA Administration. There is no penalty for prepaying a VA loan.

The veteran is liable for any losses the VA may suffer if he defaults and the house is sold in foreclosure proceedings for less than the amount of debt and sale expenses.

The loan can be assumed by either a veteran or nonveteran without the approval of the Veterans Administration. However, the original veteran is liable if the buyer defaults, unless he obtains the approval of the VA for a release of liability before the loan is assumed.

Construction Loan Mortgage

Should you decide to buy a lot and take charge of constructing your house, you would need a construction loan mortgage to finance construction.

To obtain a construction loan mortgage, you are required to submit to the lender a set of the design drawings (blueprints), specifications, and an estimate of construction costs signed by an engineer, architect, or general contractor. The lender gives these documents to its appraiser for evaluation. In addition, the lender usually requests for its legal department a recent land survey, a copy of the deed, and a statement that the property is not mortgaged. This is to ascertain that it will have a first lien to the property. Undoubtedly, the lender will search the title to ensure that there are no recorded liens to the property, but this will provide it with only a constructive notice. In addition, your statement provides the lender with an actual notice.

When your loan is approved, you will be required to provide the lender with an effective **builder's risk insurance policy** in the amount determined by the lender based on the value of construction.

Construction Schedule   Before the closing date, the lender will mail you a construction schedule that includes major areas of work such as excavation, framing, rough plumbing, etc., with each part assigned a percentage of the value of the loan. For example, if the loan is $60,000, and rough framing is assigned six percent, it means that when the rough framing is completed you get $3,600. If an item is partially completed, you get a proportion of the percentage allocated to this item.

The schedule indicates which items must be completed before you obtain the first installment (usually the foundations, rough framing, sheathing, roofing, and backfilling). When these items are completed you call the lender, which sends its appraiser or inspector to inspect the construction. The appraiser determines which items have been totally or partially completed, and takes a front-view photograph of the house for the lender's records. If everything is satisfactory to the appraiser, he informs the legal department of the amount of the first installment. Then the legal department sets the closing date, usually within a few days.

Building Loan Agreement   This is an additional document that you must sign to obtain a construction loan. It includes the description of the property, the amount of the loan, the interest rate and the method of advancing the installments. The interest is to be paid on the amount of the advanced installments only.

The loan is usually paid in four installments. The first is paid at the closing. It includes all the items that are completed in full or in part up to the appraiser's inspection. The last installment is paid after construction is completed and the certificate of electrical inspection (called **Fire Underwriters**), and the certificate of occupancy are obtained. Additionally, the grading, lawn seeding or sodding, driveways and walkways must be completed before the lender forwards the last installment to you. The other two installments are paid during progress of construction. In each case you call the lender, which sends its inspector to determine the progress of construction and the amount of the installments.

If you are building the house with FHA (Section 235) or VA financing, the property will usually be inspected during construction to see that the house is being built according to the plans and specifications originally filed with FHA or VA and that it meets the minimum government requirements. These agencies do not guarantee that the house is properly constructed in all respects, however.

Deed of Trust Mortgage

The **deed of trust** mortgage is used in several states in lieu of the mortgage document. There are three parties involved in a deed of trust: the lender, also called the **beneficiary;** the borrower, also called the **trustor;** and the **trustee.** The lender provides the borrower with the mortgage loan. In return, the borrower gives the lender a promissory note or a bond, and conveys the title to the trustee who holds it until the note is paid in full. The deed of trust is recorded in the public records.

When the note is paid in full, the lender completes a form entitled *Request For Full Reconveyance* and sends it with the promissory note to the trustee who conveys the title back to the borrower by means of a **deed of reconveyance.** The borrower must record this deed to give notice to the world that the note has been paid.

If the borrower defaults, the trustee can foreclose upon the property under the **power of sale** clause that is included in the deed of trust agreement without going through court foreclosure proceedings. After the borrower's default is established, the trustee files a notice in the public office where deeds are recorded. A waiting period of about 120 days has to elapse before the advertisement to sell the property begins. Then, the sale is advertised in the local newspapers. Finally, a public auction is held and the property is sold. The successful bidder (the buyer) gets a **trustee's deed.**

In a deed of trust, the borrower's right of redemption usually ends when the sale takes place.

Blanket Mortgage

A blanket mortgage is a mortgage secured by more than one lot. It is often used by developers to finance an entire subdivision since it is less costly than mortgaging each lot separately. When you contract a developer to build you a house, or buy only a lot from him, he must obtain a release of your lot from the blanket mortgage in order to give you a clear title. This is usually done by including a **partial release clause** in the blanket mortgage agreement by which he can obtain the release of a lot by repaying a specific portion of the debt.

Reverse Mortgage

Reverse mortgage, sometimes called **reverse annuity mortgage,** is a mortgage by which the lender makes monthly payments to the borrower. The loan is repaid in full when the borrower sells the property or from his

estate upon his death. This type of mortgage is convenient for senior citizens who need to supplement their income by utilizing the equity in their homes without having to sell them.

Open-End Mortgage

An **open-end mortgage** is a mortgage that includes a clause by which the borrower can obtain additional advances from an existing amortized loan for up to the amount that has been amortized. The new loans must be used to improve the mortgaged property. As an illustration, let us assume that the original amount of an amortized mortgage loan is $80,000, and that after 15 years $10,000 has been amortized, making the principal balance $70,000. The borrower may then open the mortgage and borrow up to $10,000 for home improvements.

Purchase Money Mortgage

A **purchase money mortgage** is a mortgage that the buyer gives the seller for a part of the price of the property. In this type of mortgage, the buyer of a property pays a portion of the price in cash and gives the seller a note secured by the property for the balance of the price. In this case, the seller becomes the mortgagee and receives the periodic payments of principal and interest. In many states, private lenders are exempt from usury laws meaning that they can charge higher interest rates than institutional lenders. On the other hand, if the interest rate on the note is too low, the Internal Revenue Service intervenes and sets a fair interest rate and taxes the seller accordingly. Some states call this type of mortgage a **purchase money deed of trust.** Purchase money mortgages can be used to finance land purchase.

Package Mortgage

A **package mortgage** includes all appliances (refrigerator, washer, dryer, etc.) that are classified as personal property in real estate mortgages. This is done to raise the value of the property, and correspondingly the amount of the mortgage loan. No item included in a package mortgage may be sold without the lender's approval.

Graduated Payment Mortgage

A **graduated payment mortgage** is a loan in which the monthly payment at the beginning of the life of the loan is less than the interest, causing the principal to increase with each payment (negative amortization). The monthly payments increase by a certain percentage each year for a specific number of years and then remain at that level until the maturity date of the loan. The Federal Housing Administration started the graduated payment loans under Section 245 of its program. Graduated payment mortgages are suited for first-time home buyers and young professionals who expect their income to increase in the years ahead.

CHAPTER 10

Financing

Most mortgage financing is provided by two major sources: the **primary mortgage market** and the **secondary mortgage market.** The primary market consists of all the institutions that lend directly to the homeowners. These lenders finance the loans they make partially from their own deposits and assets and partially by selling the loans to big institutions known as the secondary mortgage market.

In addition to the above, real estate may be purchased by **unconventional financing.** This may be defined as loans provided by financial sources other than regulated lending institutions. This includes individual lenders, city, state, and federal agencies. These sources become significant in times of tight money supply or when borrowers do not qualify for loans from regulated lenders.

Primary Mortgage Market

The primary mortgage market is comprised of all the institutions that lend directly to the homeowners. The most important lenders are:

1. Savings and loan associations
2. Commercial banks
3. Mutual saving banks
4. Insurance companies
5. Mortgage banking companies
6. Credit unions
7. Pension funds

Savings and Loan Associations Savings and loan associations (SLA) are the biggest mortgage lenders, and the best available source of construction loan mortgages for individuals. A savings and loan association must be either federally- or state-chartered. All federally- and most state-chartered associations are members of the **Federal Home Loan Bank System (FHLB),** which sets the guidelines for the lending procedures of its members. Some of FHLB requirements for loans made by member associations are: (1) the loan must be secured by a first-mortgage lien (this is one of the reasons why construction loan lenders insist that the building lot must be free from mortgages and liens), (2) loan-to-value ratio should not exceed 80 percent, (3) the mortgage loan must be amortized on a monthly basis, (4) the mortgaged property must be located within certain boundaries with respect to the SLA's home office or state, (5) the maximum term of the loan is 30 years, and (6) construction loans or second-mortgage loans can be made, but with short-term maturity.

SLAs are permitted to make **FHA-insured** and **VA-guaranteed** loans in accordance with the rules and regulations of the Federal Housing Administration and the Veterans Administration, respectively.

Most SLAs are members of the **Federal Savings and Loan Insurance Corporation (FSLIC),** which federally insures each deposit account for up to $100,000. A husband and wife can have three accounts with one association: one for the husband, another for the wife, and a third (joint) account in both names, with each account insured for up to $100,000.

Commercial Banks Although they have larger assets than savings and loan associations, commercial banks have not been as active as SLAs in mortgage lending to individuals. Commercial banks have always specialized in short-term loans to businesses, and have limited their dealings with individuals mostly to credit cards, and automobile and personal loans. The larger ones, however, have begun to offer conventional, FHA-insured, and VA-guaranteed mortgage loans. Conventional loans can be made with loan-to-value ratios of up to 80 percent for uninsured loans and 95 percent for privately insured mortgage loans.

Some commercial banks are flooded with cash and need not sell all their loans in the secondary market. These banks tend to set their own rules. Some of them offer large loans in the form of lines of credit secured by second mortgages. The loan agreement usually provides that at the end of a certain period the outstanding balance automatically becomes a conventional mortgage loan. During this period, the lender sends the borrower a monthly statement requiring payment of interest only. The borrower is allowed to repay the principal in part if he wishes to do so. However, care must be taken not to repay the outstanding balance in full, otherwise the account may be unintentionally closed. Some banks tie the interest on these loans to the prevailing prime interest rate. Some lenders offer the borrower a discount in the rate if he maintains all his savings and business accounts with them.

This type of mortgage loan is good business for the bank and convenient for the borrower. The bank charges the points on the entire line of credit up front, while the borrower may use only a portion of it. For example, if a line of credit is $50,000 and the points are 3, the bank charges the borrower up front $1500 which is $50,000 multiplied by 3 percent, while the borrower may initially borrow only $10,000. The borrower can draw and repay any amount of money within the limit of his line of credit without requiring the lender's approval.

Commercial banks must be chartered by either the federal government or the state in which their home office is located. They are members of the **Federal Deposit Insurance Corporation (FDIC),** which federally insures each account for up to $100,000. In a well publicized failure of a major bank in Illinois, the FDIC paid all the accounts in full regardless of their amount. It was a move designed to assure all depositors, particularly those of foreign countries, of the safety and security of the U. S. banking system. As with SLAs, a couple can open up three accounts at a commercial bank, and each is insured for up to $100,000.

Mutual Savings Banks Mutual savings banks are so named because they are mutually owned by their depositors. They are all state-chartered, meaning that each bank follows the regulations of the state in which it is chartered. To protect the depositors, the laws require that mutual savings banks invest their deposits in safe investments. Accordingly, they allocate about two-thirds of their investment dollars to mortgage loans. Mutual savings offer conventional as well as FHA-insured and VA-guaranteed mortgage loans. Most mutual savings banks are located in the northeastern part of the United States, mainly New York and New England.

Insurance Companies Auto, homeowner, and mortgage life insurance companies are becoming active in offering mortgage loans, mainly to their customers (mortgage life insurance is a policy whereby the insurance company pays the balance of the mortgage if either the borrower or co-borrower dies). Their activity is more in second mortgages because most of their customers already have first mortgages.

Life insurance companies are also active in home mortgages. They make conventional, FHA- and VA-loans through their branch offices, a mortgage company, or a broker.

The insurance companies' role in direct lending is limited, however. They are more active in the secondary market where they can buy big blocks of loans secured by mortgages in a single transaction.

Mortgage Banking Companies Mortgage banking companies or mortgage companies act as retailers or middlemen for large investors such as life insurance companies, pension funds, out-of-state savings and loan associations, and other big investors.

Several mortgage companies operate on a regional scale. They advertise heavily on radio and television giving a toll-free number so that prospective borrowers do not have to pay for the call, and promise to take applications by telephone and approve the loans within 24 hours.

Some of these companies merely process loans while large investors provide the money. Others have sufficient funds to finance the loans for brief periods until they sell them in the secondary market. Thus, they prefer to make loans that can be sold quickly, such as FHA-insured and VA-guaranteed loans. They also make conventional loans if they have a commitment from big investors to buy them, or if the demand for such loans is so great that they feel that they can sell them quickly if they need to.

Mortgage companies usually service the loans they make for a fee proportionate to the outstanding balance of the loan.

Credit Unions Credit unions are organized by certain groups of people or employees of particular companies as cooperatives. They specialize in consumer and home-improvement loans.

As their cash deposits increase, they channel some of this money into real estate by offering first and second mortgage loans. They are not as big as SLAs or commercial banks, but their rates are competitive, and therefore should be considered when shopping for a mortgage loan.

Pension Funds Pension fund assets are growing at very high rates. They allocate large portions of their portfolios to real estate lending. They buy big blocks of loans secured by mortgages either through mortgage companies or in the secondary mortgage market.

Secondary Mortgage Market

The secondary mortgage market consists of three major governmental or quasigovernmental agencies and one private corporation. They are:

1. The Federal National Mortgage Association (FNMA)

2. The Government National Mortgage Association (GNMA)

3. The Federal Home Loan Mortgage Corporation (FHLMC)

4. Maggie Mae (private)

The secondary market contributes significantly to stimulating construction of new housing by buying large numbers of mortgages from primary market lenders for cash. Then they package these mortgages in big blocks or pools and sell them to big investors such as pension funds, life insurance companies, mutual funds and individual investors in the form of bonds or notes secured by the mortgages. Not every individual can buy these bonds or notes directly because they are sold in large minimum denominations, but one can buy them indirectly through mutual funds.

Generally, the primary lenders continue to service the loans after they sell them. This means that they collect the payments, maintain the escrow accounts, pay the taxes, and issue the certificates of satisfaction when the loans are paid in full. In exchange for their services, lenders get a fee based on the amount of the outstanding balance of the loan. In addition, they get a portion of the discount points when the loan is initiated. The monthly payments of the principal, interest, and any prepayments are passed from lenders to secondary market purchasers. Most borrowers do not even notice that their mortgages have been sold in the secondary market, sometimes within 24 hours after the closing.

The housing industry and primary market lenders benefit enormously from the secondary market, particularly during tight money periods. By making a lot of cash available for real estate borrowing, more people can afford to buy homes and as a result more houses are built to fill the demand. Lenders benefit in two ways: (1) they make more loans on which they make more money, and (2) they can sell some loans from their portfolio at short notice should their cash reserves fall below the levels required by the federal government.

By the sheer size of their huge purchases, the secondary market institutions are able to exert great influence over the prevailing practices of mortgage lending. They are also instrumental in implementing the federal government's lending policies. Through the secondary market institutions, the federal government can set the loan-to-value and qualifying ratios (see page 77); implement using standard forms for mortgages, notes, and loan applications; and enforce the procedures by which loan applications are processed. For example, if secondary market agencies declare that the qualifying ratios should not exceed 28/36, all lenders who wish to sell mortgages to these agencies must meet these ratios. The influence of the secondary market agencies is evident in that most lenders all over the country use forms conforming to **FHLMC/FNMA** standards for mortgage loan applications, and mortgage and note documents.

The Federal National Mortgage Association FNMA, popularly known in the financial community as **Fannie Mae,** was established by the federal government in 1938 to stimulate housing construction by providing cash to government insured mortgage loans. Now, it is a quasigovernmental agency organized as a partly private, partly public corporation. Its common stocks are traded on the New York, Boston, Cincinnati, Pacific, and Philadelphia stock exchanges.

FNMA is the largest secondary market trader of mortgage loans. It buys and sells conventional, FHA-insured and VA-guaranteed mortgages through competitive periodic auctions. It finances its purchases by selling FNMA bonds and notes that are guaranteed by the federal government and secured by pools of mortgages; by selling corporate stocks; and by selling blocks of mortgage loans to pension funds, insurance companies, and other financial institutions in open competition.

Government National Mortgage Association GNMA, also known as **Ginnie Mae,** was split from FNMA in 1968. It is a division of the Department of Housing and Urban Development (HUD), and operates as a government agency. It provides assistance to housing in areas that cannot afford the prevailing interest rates through a program called **Tandem Plan.** In this program GNMA buys conventional and federally insured or guaranteed mortgage loans made to low-income families at below-market interest rates, and then sells them to FNMA or other investors at the prevailing rates with the U.S. government, through GNMA, absorbing the difference.

GNMA also has a program whereby it guarantees securities issued by primary lenders and backed by blocks of FHA-insured and VA-guaranteed mortgage loans. The investors who buy these securities are guaranteed to be paid the principal and interest on time even if the borrowers fall behind.

The Federal Home Loan Mortgage Corporation FHLMC, popularly known as **Freddie Mac,** was created in 1970 by Congress to provide a secondary market for mortgage loans.

Similar to FNMA and GNMA, Freddie Mac buys conventional, FHA-insured, and VA-guaranteed mortgage loans from lenders, then sells certificates secured by these mortgages to big investors. However, FHLMC does not guarantee the timely payments of principal and interest on these certificates.

Lenders who wish to sell their mortgages to FHLMC must use its forms for the mortgage loan application, mortgage, and promissory note.

Maggie Mae Maggie Mae is a popular name for the first nonfederal secondary mortgage market corporation. It was established in 1972 by the **Mortgage Guaranty Investment Corporation,** the parent company of the Mortgage Guaranty Insurance Corporation (**MGIC**), the biggest private mortgage loan insurer. Maggie Mae buys MGIC-insured mortgages from primary lenders, then packages and sells them to big investors.

Unconventional Financing

Unconventional financing may be defined as financing that is not provided by institutional lenders and therefore is not subject to most government lending regulations. It can be a preferred source of real estate financing in particular situations and during times of tight money. The advantage of unconventional financing sources is the elimination of a big portion of closing costs. The disadvantages are that their interest rates are likely to be higher and their terms shorter than those of conventional financing. Some examples of unconventional financing are:

Installment Contract Installment contract is sometimes called **land contract** or **contract for deed.** It is used mostly in the sale of undeveloped land or vacant lots. An installment contract may be used when a seller wants to sell a property for a certain price to an interested buyer who is unable to secure a mortgage for the entire price, but can pay the price in installments. Both parties then agree on the price, interest rate, and the number and intervals of installments. Under the contract agreement, the buyer takes possession of the property and pays the taxes and assessment charges but the seller retains the title. The deed is delivered either when the price is paid in full or in part. In the latter case, the buyer has to give the seller specific guarantees such as a **purchase money mortgage** for the balance of the price.

Before signing the contract for deed, the buyer should ensure that the seller has a **marketable title.** The deed should be held in escrow by a third party to protect the buyer if the seller dies or becomes incapacitated before transfer of title.

The other frequent application of installment contract is when an owner sells two lots to one buyer who cannot pay for both lots in cash. The seller accepts cash for one lot and installment payments for the other. The buyer can then obtain a construction loan if he wants to build a house on the lot that has been paid for in cash.

Individual Lenders Individual lenders are another source of mortgage loans. Their activity is highest in their communities and spheres of influence. They may lend directly, or through mortgage companies or brokers. Individual lenders may offer first, second, or third mortgage loans. However, they prefer second and third mortgages because this is where they can charge the most interest. In return, they may lend to less credit-worthy borrowers who might have been turned down by institutional lenders. They charge more interest, but they take more risks.

City and State Agencies City and state agencies provide mortgage loans for low-priced housing or low-income families at below-market interest rates. The funds are raised by selling tax exempt bonds. Designated local lenders are chosen to accept and process loan applications starting at a specific date on a first-come-first-served basis until the funds are exhausted. The rates are so attractive that borrowers form long lines before the lenders open their doors. As we often see on television, some borrowers line up the day before and camp out overnight to make sure that they get a loan.

Farmer's Home Administration FmHA is a branch of the U.S. Department of Agriculture. It provides mortgage loans for rural communities, by either lending directly or guaranteeing loans made by primary lenders. Eligible veterans are given preference for FmHA loans.

CHAPTER 11

Lending Practices

Lending practices are the rules that lenders follow in accepting mortgage loan applications, qualifying the applicants, determining the maximum amount of the loan, the method of charging interest, determining the discount points, deciding how the loan is to be repaid, etc. These practices change from time to time and differ from lender to lender. Nonetheless, they are guided by the fiscal and national policies of the federal government and Congress. For example, if there is a recession or the threat of one, the government is likely to issue executive orders or enact legislation to ease the conditions for mortgage lending. On the other hand, if inflation appears on the horizon, a tight fiscal policy is likely to be enforced. Each lender can set its own lending practices within government guidelines depending upon how and where it is chartered, whether it is going to sell its mortgage loans in the secondary market, and the level of its cash flow. A lender with a lot of cash is likely to offer better terms to its borrowers than a lender with little cash.

Monthly Cost of Housing (PITI)

The monthly cost of housing is composed of the principal (P), interest (I), taxes (T), and home insurance premium (I), collectively known as **PITI**. For mortgage loans with low down payments, the monthly cost of housing includes the monthly cost of the private mortgage insurance **(PMI).**

Lenders use the monthly cost of housing together with the borrower's gross monthly income to determine the maximum amount of loan for which he qualifies. Here is a description of the elements of the monthly cost of housing:

Principal The principal is the face value of the loan. It can also be identified as the amount of the obligation as stated in the promissory note, or bond. The interest in any monthly statement is calculated by multiplying the amount of the principal of the preceding month by the monthly interest rate. The principal must be reduced to zero by the end of the term of the loan.

Interest Interest is what lenders charge borrowers for using their money. A universal measure of interest is the annual interest rate, which is the percentage ratio between the amount of the yearly interest and the value of the loan. If the annual interest rate is 10 percent on a $50,000 loan, the interest per year is $5,000.

The introduction of the discount points and other service charges that have to be paid up front created two interest rates for the same loan:

1. An annual interest rate based on the nominal amount of the loan without deducting the prepaid charges. For example, if the amount of the loan is $50,000 and the annual interest rate is 10 percent, the amount of the interest per year is $5,000.

2. An effective interest rate, sometimes called **effective yield,** based on the amount of the loan less some prepaid charges. If the charges on the $50,000 loan total $2,000, the effective yield is calculated based on paying $5,000 interest on $48,000 loan. The effective interest rate calculations are based on the length of the term of the loan.

Property Taxes The monthly tax installments are calculated by dividing the annual taxes by 12. The taxes are included in the monthly cost of housing because they are obligations that must be paid.

Hazard Insurance Similar to taxes, the amount of the monthly cost of hazard insurance is obtained by dividing the annual premium by 12. Lenders require that their borrowers carry a hazard insurance policy on the mortgaged house to pay off the debt should the house burn down or be destroyed.

Lenders prefer to collect insurance installments and keep them in escrow until paid.

Private Mortgage Insurance PMI insures lenders against losses resulting from defaults on small down payment mortgage loans. Lenders, particularly those who sell their mortgage loans in the secondary mortgage market, maintain a loan-to-value ratio of 80 percent or less. Nevertheless, a loan can be made with either 10 or 5 percent down payment provided that the borrower obtains a private mortgage insurance policy for the top 20 or 25 percent of the loans, respectively. When the insured top portion of the mortgage loan is amortized or repaid, the PMI policy may be terminated with the lender's approval. The current premium of a PMI in the first year is 1/2 and 3/4 of one percent of the total value of the loan for the 10 and 5 percent down payment loans, respectively. Thereafter, the annual premium is 1/4 of one percent for either loan.

There are several private mortgage insurance companies, the most dominant of which is Mortgage Guaranty Insurance Corporation (MGIC) of Milwaukee, Wisconsin. Private mortgage insurance companies have three advantages over FHA insurance: (1) they are more competitive, (2) they insure loans of much higher value than the top limits of FHA coverage, and (3) they provide prompt service. PMI companies insure loans made only by previously approved lenders. After a lender approves a loan application, he sends the application with all pertinent documents to the PMI company for its review and final approval.

Discount Points   Discount points are charged by lenders up-front to increase the effective yield on the loan. One point is equal to one percent of the principal. Some lenders use points as a marketing technique to attract customers. For example, a lender may charge lower than the prevailing interest rate and increase the points, or vice versa.

Qualifying Ratios

The qualifying ratios are the maximum ratios between the borrower's long term monthly obligations and his gross monthly income. Lenders use these ratios to ensure that the borrowers can afford to repay the loans and to reduce the possibility of their default. The ratios are stated in the form of two figures, currently 28/36.

The first figure, 28, indicates the percentage ratio between the monthly cost of housing and the gross monthly income. Other long-term monthly obligations such as credit cards, car, furniture, and child support payments are not included in this ratio. Thus, if a borrower has $3000 gross monthly income, his monthly cost of housing should not exceed $720, which is 3000 multiplied by 28 and divided by 100.

The second figure, 36, indicates the percentage ratio between the monthly cost of housing plus all long-term monthly payments and the gross monthly income. For the same borrower of $3000 monthly income, all long-term monthly obligations should not exceed $1080 which is 3000 multiplied by 36 and divided by 100. If such a borrower has monthly installments comprised of $150 child support and $250 car installment, a total of $400,

his monthly cost of housing should not exceed $680, which is $1080 less $400. Lenders calculate the cost of housing using both ratios and use the lower figure to determine the amount of the mortgage loan.

Loan-to-Value Ratio

The loan-to-value ratio means the amount of the loan calculated as a percentage of the market value of the property. It is one of the factors that lenders use to calculate the maximum amount of a mortgage loan. The prevailing ratios are 80 percent for uninsured conventional mortgage loans and 95 percent for privately insured mortgage loans. The ratios for FHA-insured and VA-guaranteed loans are given in Chapter 9. The value of a property is determined by the lender's appraiser. If she appraises a house for $100,000 and you bought it for $110,000, you get the loan based on the $100,000 value. On the other hand, if she appraises the house at $120,000 and you bought it for $110,000, the loan is calculated based on the $110,000. A lender wants to ensure that it will get its money back should you default and should it have to sell the house through foreclosure and sale proceedings.

House Appraisal

House appraisal means determining the **market value** of the house. The market value is the maximum price for which a house can be sold in a competitive market if the house is allowed to stay on the market for a reasonable amount of time. The professional who estimates the value of the house is called the **appraiser.**

The market value of the house is determined by comparing it with other houses in the neighborhood that were recently sold and with other houses that are currently for sale. The comparison includes the size of the lot, the square footage of all the floors and the quality of construction. If the house is to be constructed (if the application is for a construction loan), the house is appraised based on the design drawings (blueprints) and specifications. For finished houses, the appraiser makes an on-site inspection of the house.

In either case, the appraiser prepares an elaborate report that includes the detailed characteristics of the house she is appraising, and the houses she is using for comparison. For houses recently sold, she reports their listing prices, the dates they were sold, the prices for which they were sold, and how many days they stayed on the market. For houses currently for sale, she reports the asking prices, and the dates they were put on the market. (She may get this information from recent issues of *Multiple Listing Services* or similar publications.) Her report includes the following:

1. The municipality, post office and zip code.

2. The style of the house (colonial, ranch, tudor, cape cod).

3. The type of construction (frame, brick, stone).

4. The type of exterior (shingle, cedar, shake, brick veneer, aluminum siding).

5. The construction of the walls of the basement (cinder blocks, concrete blocks, poured concrete).

6. The type and size of the garage (one-car garage, two-car garage, attached garage, detached garage).

7. The year in which the house was built.

8. The color of the exterior.

9. The length of the front of the lot in feet.

10. The depth of the lot in feet.

11. The property size in square feet (in case the property is irregularly shaped).

12. The square footage of all the floors. (This does not include the area of the basement unless it meets certain conditions such as having above grade windows.)

13. The number of rooms. (The hallways, foyers, and mudrooms are not counted.)

14. The number of bedrooms.

15. The number of bathrooms.

16. The number of floors and rooms on each floor.

17. The zoning district.

18. The school district.

19. The name of the elementary, junior, and senior high schools.

20. The type of insulation, if any (fiberglass, styrofoam, batts).

21. The water supply (municipal, well, private).

22. The type of plumbing pipes (copper, plastic, brass, mixed).

23. The type of roofing (asphalt shingle, shake, wood shingle, slate, tile).

24. The type of wall (drywall or sheetrock, plaster, stucco).

25. The type of heating (hot water, hot air, heat pumps, solar, steam).

26. The type of fuel (oil, gas, electric, solar).

27. The heating cost per year.

28. The amenities (alarm system, pool, tennis court, waterfront, powder room, walk to train, eat-in-kitchen, number of fireplaces, type of flooring, type of air conditioning, whirlpool bathtub, compactor, ovens).

Additionally, the appraiser takes several photos of the house she is appraising and possibly the neighboring houses. The appraiser then uses her judgment in determining the appraised value. Naturally, different appraisers give different prices.

Equity

The equity of a real property is the difference between its market value and the debt against it. If a house is worth $100,000 and the mortgage on it is $70,000, the equity in the house is $30,000. The equity changes with the rise and fall of the value of the house. If the value of the house just mentioned rises to $120,000, the equity becomes $50,000, assuming that the amount of the mortgage is unchanged. On the other hand, if the value of the house drops to $90,000, the equity is reduced to $20,000.

Repayment Plans

Mortgage loans are classified according to their repayment plans as amortized loans or balloon loans.

Amortized Loans Amortized loans are paid in monthly installments, each comprised of the interest of the preceding month plus a portion of the principal such that the principal is paid in full by the end of the term of the

loan. Most mortgage loans are of the amortized type. For fixed-rate mortgages, the monthly installments are equal throughout the life of the loan. For adjustable-rate mortgages, the monthly payment changes each time the interest rate is adjusted.

At the beginning of the life of the loan, most of the payment is used to pay the interest and only a small portion is allocated to reduce the principal. With each payment, the portion allocated to paying the interest is reduced slightly and the portion allocated to reducing the principal increases by an equal amount. By the maturity date, the balance of the principal should be reduced to zero. If not, the balance should be paid immediately.

Balloon Loans A balloon loan is a loan in which the final payment is larger than the regular monthly payment. There are two types of balloon loans: **term loan,** and **partially amortized loan.**

1. Term Loan

   In the term loan, the monthly or periodic payments consist of only the interest until the maturity date, at which time the entire loan must be paid in full. *The construction loan is a term loan.*

2. Partially Amortized Loan

   This type of loan requires the borrower to pay a specific number of amortized payments in amounts comparable to those of long-term mortgages, and a balloon payment at the end of the loan's term.

Both term and partially amortized loans have their advantages and disadvantages to the borrower. An advantage can be realized if the borrower is expecting a sum of money by the end of the loan's term from which he can pay off the loan. It is also advantageous when the borrower believes that interest rates are likely to go down substantially by the end of the term and that he can then refinance the loan at a much lower rate. The disadvantage is that the borrower has to shop for a mortgage loan when the balloon loan matures, and go through costly and time consuming closing procedures.

Usury

Usury is the interest in excess of the maximum legal interest rate as established by the law of the state. This maximum rate is adjusted periodically to reflect the prevailing interest rate. The intent of usury law is to protect borrowers from being overcharged by lenders. The penalty for usury lending varies from state to state. In some states, the violator may lose all the interest; in others it may lose the entire amount of the loan and interest. Generally, private lenders are exempt from usury limits or ceiling.

First and Second Mortgages

A **first mortgage** to a real property is a *mortgage that has no prior recorded mortgage or lien.* It is sometimes called a senior lien. A **second mortgage** is the one that is recorded after the first mortgage. It is sometimes called a **junior mortgage.**

When a first mortgage is refinanced, the second mortgage becomes a first mortgage because it was recorded before the refinancing mortgage. The refinancing mortgage becomes the second mortgage. This is the reason for the reluctance of some lenders to refinance a first mortgage if the property has a second mortgage. The priority of the mortgages may change by a **subordination clause** by which the second mortgage lender agrees that his mortgage remains a second mortgage.

Discounts

To entice borrowers, some lenders advertise low interest rates for the first few months of the life of their adjustable mortgage loans. These rates are called **discounts** or **discounted rates.** When you respond to such an advertisement, you should ask about the index and margin used in calculating the interest rate after the discount period, and how much the current interest rate would be without discount.

Discounts are sometimes used by lenders to qualify borrowers who may not qualify if the normal rates were used. If this is the case, you should determine before you accept such an offer whether you will be able to afford the payments after the discount expires.

Buy-Down

Buy-down is a technique used by developers to attract buyers. A developer makes an agreement with a lender whereby he pays the lender a sum of money to compensate it for offering the people who buy houses from him lower than the **prevailing interest rates** (and consequently lower payments) for an early period of the mortgage term. This way the houses sell faster because more buyers can qualify to buy them. It is doubtful, however, that buy-down will save the buyers any money because the developer is likely to raise the prices of the houses to compensate for the buy-down he pays to the lender.

This method is advantageous to the buyer if he is expecting his income to rise by the end of the discount period. However, before you decide to buy a house under a buy-down plan, make sure that you can afford the regular payments after the discount rate expires.

Certificate of Reduction of Mortgage

This certificate is needed when a buyer is assuming the seller's mortgage. The buyer requires a proof of the exact amount of the reduced mortgage or the outstanding principal that he will be obliged to pay. The contract of sale should contain a clause that the seller delivers to the buyer a **certificate of reduction of mortgage** executed by the **lender.** The certificate must describe the mortgage and state that the lender is the holder of the mortgage loan. It should also indicate the outstanding principal at a specific date, usually the closing date, the interest rate, the amount of the monthly installments, and that the lender knows the property is being sold and the mortgage assumed. The certificate must be signed by the lender or mortgage holder, who must also have it notarized if it is to be recorded.

Estoppel Certificate

This certificate is also required if the buyer is assuming the existing mortgage. It is prepared by the lender but signed by the **seller** and states the amount of the outstanding principal, the interest rate, and the monthly pay-

ment of the assumed loan. The purpose of this certificate is to have the seller's consent to the lender's statement.

Shopping for a Mortgage Loan

The first step toward obtaining a mortgage or construction loan is to call as many lenders as you can and ask about their interest rate, points, terms of the loan, and other requirements for both fixed- and adjustable-rate mortgages. Get loan applications from the two or three lenders that offer the most favorable terms. Apply to only one or two lenders; filling out mortgage applications is a time consuming process. It is also costly, because lenders require a nonrefundable appraisal fee (a few hundred dollars) with each application.

Mortgage Application Forms

Most lenders use a printed mortgage loan application conforming to the standards of the **Federal National Mortgage Association/Federal Home Loan Mortgage Corporation (FNMA/ FHLMC).** The same form is used for construction loan applications. The main items on this form are:

1. The type of mortgage: conventional, FHA-insured or VA-guaranteed.

2. The address of the property.

3. The legal description of the property (attach a sheet if necessary).

4. The name and telephone number of the person to call regarding the application (usually the borrower or the co-borrower).

5. The purpose of the loan: purchase (if you are buying a finished house); construction (if you have bought a lot and want to build); construction–permanent (if you have bought a lot and wish to have a construction loan and a long-term loan closed in one session); refinance (if you are refinancing a construction or an existing long-term loan).

6. Items to be completed only if the application is for a construction or construction–permanent loan. They include the lot's original cost, present value, year acquired, and cost of improvements (cost of building the house).

7. The name under which the title will be held.

8. Source of down payment and closing or settlement charges.

9. The borrower's and co-borrower's names; current addresses; names and addresses of their present employers; types of business; social security numbers; home and business telephone numbers.

10. Gross monthly income of the borrower and co-borrower.

11. Monthly housing expenses.

12. Legal and financial questions for the borrower and co-borrower:
    a. Have you any outstanding judgment?
    b. In the last seven years, have you been declared bankrupt?
    c. Have you had property foreclosed upon or given title or deed in lieu thereof?
    d. Are you a co-maker or endorser of a note?
    e. Are you a party in a lawsuit?
    f. Are you obliged to pay alimony, child support, or separate maintenance?
    g. Is any part of the down payment borrowed?

    An explanation is required if a "yes" answer is given to any of the above questions.

13. A schedule of the borrower's and co-borrower's assets.

14. A schedule of the borrower's and co-borrower's liabilities and pledged assets.

15. A schedule of the real estate owned by the borrower and co-borrower.

16. A list of previous credit references.

17. An agreement that:
    a. The mortgage loan is to be secured by a first mortgage or deed of trust to the property.
    b. The property will not be used for any illegal or restricted purposes.
    c. All statements made in the application are true.
    d. Verification may be obtained from any source named in the application.
    e. The borrower and co-borrower should indicate whether they intend or do not intend to occupy the property.

18. Authorization to lender to request a credit report on the borrower and co-borrower.

19. Information for government monitoring purposes. In order to monitor the lender's compliance with **equal credit opportunity and fair housing laws** the federal government encourages the borrower and co-borrower to furnish information regarding their race, national origin, and sex. If the borrower or co-borrower chooses not to furnish such information, federal regulations require the lender to note race and sex on the basis of visual observation or surname.

20. The borrower and co-borrower must sign the application and indicate the date of signing.

Verification of Employment Lenders mail their loan applicants a "Request For Verification Of Employment," to be filled out by the borrower's and co-borrower's current employers and mailed directly to the lender. The verification of employment includes date of employment, present position, probability of continued employment, base pay, overtime, commission, bonus, and remarks.

Verification Of Deposits The lender mails you a "Request For Verification Of Deposit," to be filled out by your depositories (the banks in which you have deposits), and mailed directly to the lender. Verification includes the type of account, account number, current balance, the date it was opened, and the average balance for the previous two months.

After receiving a mortgage loan application, the lender assigns an appraiser to determine the costs of construction or the market value of the house.

Loan Approval

When your mortgage loan application is approved, the lender mails you, the borrower, a letter to that effect. The letter states the terms and conditions of the loan, the cost of the closing or settlement and the duration of the commitment (from 15 days to a few months). The lender also requests a copy of the deed, a recent land survey, and a signed copy of the terms and conditions to indicate your consent. At this stage, the lender may demand part or all of the discount points.

Borrower's Right to Rescission

Under federal law, a borrower has the **right to rescind** (cancel or repeal) a loan transaction that may result in a lien, mortgage or other security inter-est on **his home.** You can exercise this right when you refinance a construction or a long-term mortgage loan or when you apply for a second mortgage provided that you are living in the house. However, this right does not apply when the loan is used to buy a previously owned house.

When applicable, the lender is obliged to mail you a "Notice Of Right Of Rescission" after it approves your loan application and before the closing date. You actually get two notices, one in your name and the other in the co-borrower's name. The notice states that you have the legal right under federal law to cancel this transaction if you desire to do so, without any penalty or obligation, within three business days from the date of the issue of the notice or any later date on which all material disclosures required under the **Truth in Lending Act** have been given to you. If you cancel the transaction, any lien or mortgage arising from the transaction is automatically void. You are also entitled to receive a refund of any down payment or other consideration if you cancel. If you decide to cancel the transaction, you may do so by notifying the lender by mail, telegram or any other form of written notice not later than midnight of the date stated in the notice (based on three days notice). For your protection, the use of registered or certified mail with return receipt is recommended.

When you exercise your right to rescind, the lender is obliged to return to you any money or property given as earnest money, down payment or otherwise, within 10 days after receiving your notice. The lender shall also take the necessary action to terminate any security interest (such as liens) created under this transaction.

CHAPTER 12

Insurance and Surety Bonds

S everal kinds of insurance are required in connection with the construction of your house. Which ones you will need depends on how construction is going to be managed. If you buy a lot and hire contractors to do the excavation, plumbing, rough framing etc., you must buy workmen's compensation, disability benefits, builder's risk, and family liability insurance policies. But if you buy a lot and hire a general contractor to build the entire house, your contract with him should state that he carries the first three insurances and that he must present you with a copy of the policies before construction starts. In all cases, you must be covered by a family liability insurance.

Workmen's Compensation Insurance

Workmen's compensation insurance provides cash benefits and medical care for workers who become disabled because of an injury or sickness **related to their work.** Some states allow this type of insurance to be issued only by the **State Insurance Fund,** which is owned by the state. In other states, this insurance is provided solely by private insurance companies. In the remaining states, either the State Fund or private companies may issue this type of policy.

The benefits under this insurance are only those provided by law and approved by the state's **workers' compensation board.** If death results, benefits are payable to the surviving spouse and the dependents as defined by law.

In addition to the benefits paid to the injured, workmen's compensation insurance provides the insured with **legal defense** and pays any **court award** in jurisdictions where an injured worker can bring a lawsuit against his employer.

If you are managing the construction of your house without having a workmen's compensation insurance coverage, you are personally liable for any compensation benefits due to injured worker(s), and to his (their) family and dependents in the event of his (their) death. In addition, you may be subject to penalties in accordance with the law. Such liabilities can be of a staggering magnitude and can exceed by far the total value of your house. The basic benefits the workers are entitled to, and the disability classifications will be reviewed for your information. The details differ from state to state, but the basics are similar.

Medical Care   In almost all states, the injured worker who is eligible for worker's compensation is entitled to all necessary medical care as the nature of the injury or the process of recovery may require.

Medical treatment may include, depending on state laws, medical, osteopathic, dental, podiatric and chiropractic treatment, surgery and hospital care, X-rays, laboratory tests, prescribed drugs, authorized nursing service, and any medical or surgical appliances required by the injury. The worker may choose any physician, podiatrist or chiropractor who is approved by the worker's compensation board.

Cash Compensation   Workers who are totally or partially disabled (disability classifications are explained below) and who are eligible for cash benefits, receive two thirds of their average weekly wage, but no more than the maximum benefit. The average weekly wage is based on payroll records for the year prior to the date of the disability or accident.

There are two benefit maximums, one for partial disability and the other for total disability. The maximum weekly benefits for any accident or sickness are periodically adjusted up-

ward and are substantially different for the partial and total disabilities.

In the Event of Death   If the worker dies from a compensable injury, the surviving spouse and dependents as defined by law are entitled to weekly cash benefits. The amount is figured as two thirds of the deceased worker's average weekly wage for the year before the accident, but in no event may the compensation exceed the established benefit maximum, no matter how many dependents are involved.

If there is no surviving spouse or children, other dependents, such as parents, grandchildren, or brothers or sisters, as defined by law, may be entitled to cash benefits if dependency is proven.

In addition to these benefits, funeral expenses are payable up to a certain amount.

Disability Classifications   The number and the details of disability classifications differ from state to state. Here are the five basic classifications:

1. **Temporary Partial Disability.** The wage-earning capacity is only partially lost, and on a temporary basis.

2. **Temporary Total Disability.** The injured worker's wage-earning capacity is totally lost, but only on a temporary basis.

3. **Permanent Partial Disability.** Part of the employee's wage-earning capacity has been permanently lost. Benefits are payable as long as the partial disability exists.

4. **Permanent Total Disability.** The employee has permanently and totally lost wage-earning capacity for his job. There is no limit on the number of weeks payable.

5. **Disfigurement.** Serious and permanent disfigurement of the face, head, or neck may entitle the worker to compensation up to a maximum amount.

Hearings and Appeals An employee who applies for compensation under a workmen's compensation policy forfeits his right to sue his employer for damages. An injured employee may choose to reject the compensation provided by law and instead sue his employer. However, the conditions for rejection of the law by a worker are extremely difficult. In some states, a worker may not sue his employer. If the injured worker is not satisfied with the award as specified by law, he can appeal before a court designated by law.

If a claim is challenged, the administrative authority holds hearings before a worker's compensation judge. The judge takes testimony, reviews medical and other evidence and decides whether the claimant is entitled to benefits. If the claim is found compensable, the judge determines the amount and duration of the compensation award.

Either side may appeal the judge's decision by applying for a board review in writing within a specific number of days. If the application is granted, the board assigns a panel of three members to review the case. This panel may affirm, modify, or rescind the judge's decision or restore the case to a judge for further consideration. In the event the panel is not unanimous, any party may apply in writing for a full board review, and the full board must review the decision and either affirm, modify, or rescind the panel's decision.

Workmen's Compensation Insurance Premium Premiums for workmen's compensation insurance are computed by multiplying each worker's wages, up to a certain dollar figure, by the rate specified for his or her work classification (carpenter, plumber, mason, etc.) taking into consideration the volume of construction.

When buying a workmen's insurance policy, you are required to give the insurance representative an estimate of the number and classification of the workers you intend to employ directly and how long their assignments are expected to last. This is in order to estimate the premium. Even if you plan to assign all items to contractors who carry workmen's compensation insurance, you still have to pay a deposit the value of which depends on the value of construction. The policy is effective for one year, renewable. At the end of the year, an insurance representative pays you a visit to check your records to determine the number and classification of the workers you have hired directly. If you gave all or a part of the work to contractors, you are required to present the names and addresses of these contractors and a copy of their contracts or bills to him as evidence. Based on this information, he estimates the insurance premium for the preceding year. Accordingly, you might have to pay more money or may get a refund.

Disability Benefit Insurance

Disability benefit insurance provides your workers with protection for disabilities incurred through accidents or diseases **not related to their work.** This insurance can be bought from the state insurance fund or private insurance carriers, depending on the state. Whoever manages construction should carry this insurance (the premium is small).

Disability insurance usually pays about the same as unemployment benefits. All state plans provide for maximum weekly benefits and a maximum number of weeks for paying benefits. A waiting period of one week must elapse before benefits begin. Some states pay this week retroactively if the illness lasts more than a specific number of weeks.

Builder's Risk Insurance

Builder's risk insurance provides protection against losses caused by fire, lightning, windstorm and vandalism. During construction, the wood framing is exposed and this makes it vulnerable particularly at night or during weekends when the house is unattended. This is why construction loan lenders require that you, the borrower, must carry this insurance before the closing in an amount equal to the appraised value of construction. Some lenders require that you pay one year's premium by the closing date, others permit you to pay for it in installments. This insurance usually covers the structure and the materials to be used in construction (such as a pile of lumber) that are adjacent to it. However, records, documents, drawings and specifications are not covered under this insurance.

The amount of the premium depends on the face value of the policy, the type of construction, and the proximity of fire fighting facilities such as the distance to the nearest fire station or fire hydrant.

Other coverage such as theft or robbery may be added to the builder's risk policy. Such coverage is expensive, however.

The builder's risk policy is terminated when the house is completed and the certificate of occupancy is obtained. It is to be replaced by a hazard or homeowner's insurance policy.

Family Liability Insurance

As soon as you buy a lot, make sure that you are covered by a **family liability insurance.** If you are living in your own house, your home insurance policy may provide you with liability coverage. Check your policy and consult your insurance agent. But if you are living in an apartment, you may not have such coverage. Some insurance companies offer policies that combine theft and fire insurance for furniture and personal belongings in the apartment with liability insurance.

The reason for carrying such insurance is to cover you in case somebody trespasses on your property and gets hurt and sues you whether or not it is "your fault." A construction worker may be injured and decide to sue you for negligence, even though you might not be negligent. There are two legal rules in this regard: (1) everybody has the right to sue, and (2) you can be sued even if you are innocent.

The liability insurance company pays for the **legal defense,** and the **judgments** against the insured, up to the face value of the insurance, each occurrence. If a person who is covered by liability insurance suffers a loss involving a land vehicle or watercraft, the liability insurance pays only after all other applicable insurances have been exhausted. It is therefore recommended that you buy liability insurance from your auto or boat insurance company so that you would be dealing with only one company. Such a company must be reputable and reliable.

Homeowner's Policy

As soon as you obtain the certificate of occupancy and move into the house, you should replace the builder's risk policy with a homeowner's insurance policy. Homeowner's policies are not all the same. Thus, you must buy a policy from a reputable company to ensure that your house is protected. A good homeowner's policy provides the following coverage:

1. For physical loss to the house up to the face value of the policy. There are upgraded policies whereby the insurance company increases the coverage each year according to a specific index to reflect building cost increases. In these policies, the replacement of the property is guaranteed. The first $200 or so is deductible.

2. For personal property for an amount equal to 50 percent of the coverage of the house. The first $200 or so is deductible.

3. For family liability protection, usually a few hundred thousand dollars, each occurrence.

4. For guest medical protection, about $1000 each person.

The policy explains in detail what and who is covered and what and who is not covered under each item. There-

fore, you should read the policy carefully. It is important to know what is *not* covered in your homeowner's policy. It is a general rule that the following items are not covered by homeowner's insurance policies:

1. Water damage caused by flooding, backups of sewers or drains, and subsurface or underground water.

2. Earth movement resulting from earthquakes or other reasons, volcanic eruptions, landslides, mudflows, erosions, or rising, sinking, shifting, expanding or contracting of the earth.

3. Enforcement of any ordinance or law regulating the construction, repair, or demolition of buildings or other structures.

4. Neglect by an insured person to take all reasonable steps to save and preserve the property at and after a loss or when the property is endangered by a loss covered by the policy. (This means that you must promptly call the fire department if there is a fire!)

5. Nuclear reactions, including radiation and radioactive contamination.

6. War or warlike acts.

7. Freezing of plumbing, heating, or air conditioning system.

8. Settling, cracking, shrinking, bulging, or expansion of foundations, walls, floors, roofs, ceilings, pavements, or patios.

9. Seepage or leakage of water or steam over a period of time from the plumbing, heating, or air conditioning system.

10. Vandalism if the house covered had been vacant for more than 30 consecutive days immediately prior to the loss.

Surety Bonds

Surety bonds are provided by companies that specialize in them. A **surety** is a party that assumes liability if another party fails to perform an agreement, fulfill a contract, or pay a debt. A surety bond is a written agreement that states the terms and obligations under such an agreement. In the course of building your house, one or more of the following bonds may be used:

Performance Bond   If you assign the entire house to a general contractor, the successful completion of the house depends on him. If he quits in the middle of the work or if he goes bankrupt, the project will come to a halt. Therefore, you may want to protect yourself against such an occurrence by demanding that he provides you with a performance bond. The bond guarantees that the house will be built as agreed upon in the contract and according to the drawings and specifications. Generally, the bond covers the warranty period stated in the contract, which should be at least one year after the date of the certificate of occupancy. The bond should cover the changes you may require during construction. The face value of this bond should be higher than the amount of the contract.

Payment Bond   This bond guarantees payment of workers and material suppliers. It protects you against claims by these individuals. The face value of this bond is about 50 percent of the value of the contract since wages represent only a part of construction costs.

When you require the contractor to provide you with surety bonds, you have to keep in mind that he is going to pass the premium charges plus a mark-up to cover his overhead expenses on to you.

Bond to Discharge a Mechanic's Lien   A surety bond in the amount claimed by the workers or material suppliers can be used to discharge a mechanic's lien. The lien holder may then legally attack the surety bond to satisfy his claim.

Default by the Contractor   If a bonded contractor defaults, the surety discharges its obligation under the bond agreement up to the face value of the bond. It is up to the surety company to decide how to complete the contract. If the contractor goes out of business, the surety company will have to assign the remaining work to another contractor either through competitive bidding or otherwise. If the contractor's failure is a result of low bidding and he has stopped to cut his losses, the surety company may provide him with financing to complete the job.

CHAPTER 13

Closing and Closing Costs

Closing or **settlement** of a real estate transaction is the process by which the promises and agreements made by the parties are fulfilled or executed. In most sale transactions, two closings take place: (1) the closing of the buyer's loan and the payment of the existing seller's mortgage, if any, and (2) the closing of the sale of the house or lot, whereby the seller delivers the agreed-upon deed to the buyer and the buyer delivers the agreed-upon price to the seller.

After the transaction takes place, all relevant documents **must be recorded in the proper sequence.** The satisfaction of the seller's mortgage, if any, must be recorded first, followed by the buyer's deed, and then the buyer's mortgage.

Closing Meeting

The closing meeting may be held at the office of either the lender of the new mortgage loan, the seller's lawyer, ·the buyer's lawyer or the real estate agent, depending on the circumstances of the particular closing. For instance, construction loan closing is likely to be held in the lender's office because there is no real estate agent or seller involved.

Present at the closing meeting are the seller and the buyer; each is usually accompanied by his or her lawyer. Also present is the representative of the title search or title insurance company. If the transaction involves obtaining a new mortgage loan and paying off an existing one, a lawyer or other representative of the lender who made the new loan must be present. A representative of the lender of the paid-off loan may or may not be present. If he is not, the representative of the title search or title insurance company is usually entrusted with delivering the check of the paid-off loan to the office of that lender. In return, he gets a gratuity from the buyer. The real estate agent or agents who brought the buyer and seller together may or may not be present.

Closing Agent

The closing agent is the person who conducts the closing or settlement proceedings. She is usually a representative from the office in which the closing is taking place. The closing agent is charged with calculating the closing costs, as well as the division of taxes, the dollar-value of heating oil in the tank, and other charges between the seller and the buyer; disbursing funds to the seller and title insurance company; calculating and withholding the property taxes and insurance premiums in an escrow account; and ensuring that the documents are signed and acknowledged in the proper sequence.

Escrow Closing

In escrow closing, there may not be a meeting between the buyer and the seller. Rather, they both choose a neutral third party called an **escrow agent** or escrow holder to handle the transaction. The escrow agent may be a lawyer, the escrow department of a bank, or the title insurance company. A real estate broker may be an escrow agent, but not for a transaction on which she is earning a commission, so as to avoid having a conflict of interest. Many states require that the escrow agent be licensed and bonded, because of the considerable amount of valuables she may be entrusted with.

The buyer and seller choose the escrow agent at the time of signing the contract of sale. They then sign an **escrow agreement** in which they state in writing the sequence of actions to be performed by each party. The first step would be for the seller to sign and deliver the deed to the escrow agent with instructions not to deliver it to the buyer until the agreed upon price is received. He also delivers to the agent the tax schedule, recent survey, certificate of occupancy, and all papers pertinent to the transaction. The buyer delivers a deposit or down payment to the agent with instructions to keep it in escrow and to obtain a title search or title insurance policy. If the property has an outstanding mortgage loan, the escrow agent contacts the lender notifying it that the property is being sold, and requests a certificate of reduction of mortgage. She also examines the certificate of title, or title insurance policy. If she finds the title to be marketable, she demands the balance of the price from the buyer, delivers the deed to him, and the price to the seller. If on the other hand, she finds the deed to be defective, the escrow agent returns the down payment to the buyer and the deed to the seller.

Real Estate Settlement Procedure Act

Real estate settlement procedure act **(RESPA)** was passed by Congress in the mid-1970s to protect mortgage loan borrowers. The act applies to the lender whose deposits are insured by FSLIC or FDIC; to loans that are to be sold to FNMA, GNMA or FHLMC; and to loans that are FHA-insured or VA-guaranteed. In practice, institutional lenders mail good faith cost estimates to all their applicants.

When you submit a mortgage loan application that meets RESPA requirements, the lender must provide you with the following:

I Information Booklet The lender must mail you an information booklet,

Settlement Costs, prepared by the Department of Housing and Urban Development (HUD) no later than three business days after your application is filed. The booklet is in two parts: part one describes the settlement process. It also contains information on your rights and the remedies available under RESPA and alerts you to unfair or illegal practices; part two is an item-by-item explanation of settlement or closing services and costs with sample work sheets to help you make cost comparisons among various lenders.

II Good Faith Estimate of Settlement Costs The lender must also mail you no later than three business days after your application is filed a **good faith estimate** of the settlement costs you are likely to incur. The lender should itemize the cost of each item and the total cost of all items.

III Uniform Settlement Statement This is a form on which the services provided to you are itemized and the fee for each item is indicated. This statement (also called **closing statement**) must be prepared by the closing agent before the closing or settlement meeting. You, the borrower, have the right to inspect this statement **one business day before the closing.** The closing agent may not have all costs available the day before the closing but he is obliged to show you, upon request, what is available. In the case of escrow closing, the statement will be mailed to you soon after the closing. The statement form was developed by the Department of Housing and Urban Development (HUD).

The uniform settlement statement is not used in situations where the lender charges a fixed amount for all closing costs. In this case, the borrower must be informed of this fixed amount at the time of loan application. Additionally, the lender is required to provide the borrower with a list of the services to be rendered, within three business days after the application is filed.

IV Truth-in-Lending Under the federal law called the **Truth-in-Lending Act** (also known as **Federal Reserve Regulation Z**), the lender is required to provide you with a **truth-in-lending statement** by the time of loan completion. It discloses the annual interest rate and the effective interest rate or effective yield. The statement also includes the prepaid finance charges such as appraisal fees, closing charges, interest to the end of the month, processing and commitment fees, and inspection fees for construction loans. Also included are the penalty charges for late payments (between four and five percent of the overdue amount); the interest rate to be charged if the loan is in default; and the penalties, if any, for prepayment of the mortgage in part or in full.

Escrow Accounts In most mortgage loans, lenders collect with each payment approximately one-twelfth of the estimated annual property taxes, assessments, insurance premiums and other recurring charges. They keep these funds in a separate account called **escrow account** or **reserve account** from which they pay taxes and insurance premiums when they are due.

At the closing, the borrower is required to make an initial deposit into the escrow account to ensure that there are enough funds to pay for the taxes or insurance installments. RESPA limits the maximum amount a lender can require a borrower to deposit into an escrow account at the closing to:

1. An amount that would be sufficient to pay taxes and insurance premiums.

2. An additional amount not in excess of two months' payments of the estimated taxes and insurance premiums.

Lenders revise monthly tax and insurance withholdings each year to reflect any changes in these items.

Protection against Unfair Practices

The law provides consumers with protection from unnecessarily high settlement charges due to abusive practices. Some unlawful activities are:

Kickbacks The law prohibits anyone from giving or taking *a fee, kickback, or anything of value* as part of an agreement that business will be referred to a specific person or organization. It is also illegal to charge or accept a fee or part of a fee where no service has actually been performed.

The prohibition is aimed at eliminating arrangements in which one party agrees to pay a part of his fee to a referring party in exchange for obtaining business. This could lead the party providing the services to inflate his fee to cover the payments to the referring party, resulting in higher costs to the borrower. There are criminal penalties of both fine and imprisonment for any violation of this law. There are also provisions for the borrower to recover three times the amount of the kickback, rebate or referral fee involved, through a private lawsuit. If the borrower wins the lawsuit, the court may award him court costs plus lawyer's fees.

Title Companies Under the law, the seller may not require, as a condition of sale, that the buyer purchase a title insurance from a particular title company. The violator can be held liable to the buyer in an amount equal to three times all charges made for the title insurance.

Fair Credit Reporting There are nationwide credit reporting companies whose business is to compile credit data on individuals. The data includes all outstanding debts such as mortgages, personal loans, car loans, credit cards, and department store charge accounts; whether you pay your bills on time; whether you have filed for bankruptcy; whether you have been sued; and any judgments against you.

The **Fair Credit Reporting Act** does not give you the right to inspect or physically handle your actual report at the credit reporting company, nor to receive an exact copy of the report. You may at any time obtain a summary of your credit report for a small fee (about $10). However, you may obtain the summary report free of charge if you are denied a mortgage or a personal or auto loan due to a bad credit report, by mailing a request to the credit company naming the lender and the loan number. It is very possible that the report is not completely accurate and you have the right to challenge its accuracy and request that a correction be made.

Equal Credit Opportunity The Equal Credit Opportunity Act prohibits lenders from discriminating against credit applicants on the basis of **race, color, religion, national origin, sex, marital status, or age** (provided that the applicant has the capacity to enter into a binding contract) or because all or part of the applicant's income derives from any public assistance program. If you feel you have been discriminated against by a lender, you may ask the lender about the identity of the Federal Agency that administers compliance with this law and consult with that agency. You also have the right to sue the lender.

Closing Costs

Closing or settlement costs are expenses paid for services in association with real estate transactions. These costs can be substantial and the buyer must have the funds to pay for them before the closing. The following are possible settlement services for which you may be charged:

Origination Fee   Origination fee is what some lenders charge their borrowers to cover their costs and expenses for the services associated with processing the mortgage loan. Such expenses include reviewing the loan application form, searching the title, appraising the property, obtaining and reviewing a credit report on the borrower, preparing the note and mortgage documents, overhead expenses, etc.

Discount Points   As explained in Chapter 11, discount points are charged by lenders when the loan is initiated to increase their yield on the loan.

Appraisal Fee   This fee is to compensate the lender for the costs of hiring an independent appraiser or assigning a member of its staff to determine the value of the mortgaged property. Some lenders charge a flat appraisal fee, others charge higher fees for high priced houses. Most lenders require that a check for the appraisal fee accompany the loan application on a nonrefundable basis, otherwise they will not process the loan.

Credit Report Fee   This fee covers the cost of obtaining a credit report on you, the borrower. The lender uses this report to learn about your outstanding loans and other financial obligations in order to calculate the maximum mortgage loan you are eligible for. The lender also wants to know if you pay your bills on time so as to determine whether or not you are creditworthy. This fee is small.

Lender's Inspection Fee   This fee is often associated with construction mortgage loans. It covers the costs of inspecting construction each time you request a loan installment. This inspection may be made by a lender's employee or by an independent inspector.

Private Mortgage Insurance Application Fee   This covers processing the application for a private mortgage insurance policy, which is required for small down payment mortgage loans.

Assumption Fee   This fee is charged to cover the processing costs if the buyer assumes the balance of the existing mortgage.

Interest   Lenders require that borrowers pay at the closing the interest that accrues on the loan (or the first installment for construction loans) from the date of the settlement to the beginning of the following month. For example, if the closing date is October 10, your first monthly payment will be due on December 1 and covers the interest for the month of November. The interest on the loan for the period from October 10 to November 1 will be collected at the closing.

Mortgage Insurance Premium This type of insurance protects the lender from losses should the borrower default on the loan. The lender may require you to pay the premium before or on the day of settlement. The yearly premium may be paid in one or several installments. This insurance is not always required and should not be confused with mortgage life insurance, which pays off the balance of the mortgage if the borrower or co-borrower dies.

Hazard Insurance Premium Lenders require the borrower to have an effective hazard or homeowner's insurance policy before the closing. The face value of the policy is proportional to the value of the completed house (usually 80 percent), regardless of the amount of the mortgage loan. As an illustration, if a house is appraised at $200,000 and the mortgage loan is only $50,000, the lender may require a hazard insurance policy for at least $160,000, which is 80 percent of the $200,000. This insurance protects you, the owner, and the lender against loss due to fire, windstorms, and other natural hazards. The protection under this policy does not include losses caused by flooding in flood-prone areas.

Closing Fee The closing fee, sometimes called settlement fee or closing charge, is to cover the services of the settlement agent. This fee may be paid by either the buyer or the seller and this should be negotiated between both at the time of signing the contract for sale.

Abstract or Title Search These charges cover the costs of searching the public records and preparing the abstract of title. The borrower or the seller or both may pay for these charges depending on the local customs. The contract of sale should indicate who pays for these charges.

Mortgage Title Insurance This is a one-time fee or premium for the mortgage title insurance to be paid at the closing. It is customary that the buyer pays for this insurance unless the seller agrees in the contract of sale to pay part or all of it.

Attorney's Fees You may be required to pay for legal services provided by the lender in examining the legal documents pertaining to the title and other matters. Usually, this fee is charged if you do not have a lawyer.

Owner's Title Insurance This fee is for owner's title insurance protection. When a mortgage title insurance is provided, the title insurance company is required to offer the buyer an owner's insurance policy at a slight additional cost. In some areas the buyer pays for this insurance, in other areas, the seller provides the buyer with this policy.

Recording Fees These charges are usually paid by the buyer (borrower). They include the deed recording fee, mortgage recording fee, and building loan filing fee.

Government Taxes   These taxes vary from state to state. Usually the state regulates who pays which taxes. These taxes are:

1. **State mortgage tax.** This is paid for by the borrower or shared between the borrower and lender.

2. **State tax or stamps.** Usually paid for by the seller.

3. **City and county tax or stamps.** Usually paid for by the seller.

Survey   The lender or the title insurance company requires a recent land survey showing the boundaries of the lot and the exact location of the house.

Home Inspection   This inspection is for previously owned homes. The buyer usually pays for such inspection.

Adjustments between Buyer and Seller

The adjustment between buyer and seller involves dividing the taxes, value of remaining heating fuel, and water charges between the buyer and seller.

Property Taxes   Property taxes can be a big item. In some expensive areas, taxes can be well over $10,000 per year. As an illustration of tax adjustment, let us assume that the whole year's taxes are payable on June 30 of each year to cover until June 30 of the following year. Let's say that the closing occurs on November 30. In the adjustment, you have to pay the seller a portion of the taxes that covers the period from November 30 to the following June 30, because he has already paid for this period.

Fuel   When you buy a previously owned house heated by oil, it is normal to have some oil left in the tank when the transaction takes place. You are required to pay for this oil based on the number of gallons in the tank multiplied by the prevailing price per gallon.

CHAPTER 14

Architectural Styles

An important feature of any house is its **architectural style,** which is its look as seen from the outside. The major components of an architectural style are the shape of the exterior walls; the style of the roof, windows and main entrance; the building materials of the exterior siding and roofing; and the decorative detailing.

The Shape of the Exterior Walls

The shape of the exterior walls is a function of the floor plans and the front elevations.

The Floor Plan The floor plan is the arrangement of the bedrooms, living room, family room, dining room, kitchen, bathrooms, foyer, laundry room, etc. looking at them from above after removing the roof. The simplest form of a floor plan is the rectangle. Many styles have rectangular floor plans: the Cape Cods, most of the Colonials, and some Ranches. The next simplest form is the L or T shape. It is formed by a rectangle having a square

or rectangle attached to its side. Among the styles that have L or T shaped floor plans are the Split-Levels and some Ranches. A third floor plan is the U shape. This can be seen in some Ranches. Not all floor plans are that simple, however. Some plans have irregular shapes, as in the case of some Tudors, Victorians, and Contemporaries.

Front Elevations The front elevations are the four sides of the house—the front, the back and the two sides—as seen by a person facing each side. The elevations of the exterior walls reveal the number of stories, the width and depth of the house. A one story house indicates that it may be a Ranch or a Cape Cod. A one and a half story house may be a Dutch Colonial, Cape Cod, French Colonial or French Provincial. A two or three story house may be a Georgian, Tudor, Mediterranean, or Mission. The elevations also show whether the front of the house is symmetrical or not, which is one of the identifying features of several styles.

Roof Style

The roof style is the second principal factor in shaping and identifying the style of a house. Certain house styles are identified with specific roof styles. For example, the Dutch Colonial is identified with the gambrel roof; the French Provincial with the steep hip roof; the Tudor with the steep cross gable roof; the Shed with the shed roof; and the Mansard with the mansard roof. The style of a roof includes its shape, pitch (meaning the slope of its sides), the shape and size of the eave overhangs, and the shape and size of the dormers, if any.

Roof Shapes The **Gable** is the most common shape. If the main entrance door of the house faces the sloped side of the roof, it is identified as **Side Gable;** if it faces the triangular side, it is defined as **Front Gable.** If the roof is L shaped, it is called **Front and Wing Gable;** if it is T shaped, it is called

Cross Gable. These gable type roofs are illustrated in Figure 14.1.

The other most easily recognized roof shapes are the **Gambrel** with narrow or flared eave overhangs, **Hip, Shed, Mansard, Salt Box** and **Flat,** which are shown in Figure 14.2.

Roof Pitch The roof pitch is the angle between the sloping sides of the roof and the horizontal. If this angle is less than 30°, the roof is classified as **low pitch;** if the angle is between 30° and 45°, it is classified as **normal pitch;** and if it is more than 45°, it is classified as **steep pitch** (Figure 14.3). Changing the roof pitch can have a dramatic effect on the style of the house. For example, the steep pitch hip roof is one of the main characteristics of the French Provincial style, while the low pitch hip is commonly found in the Mediterranean style and some Ranches; the steep pitch gable is a characteristic of the Tudor, while the normal pitch gable is

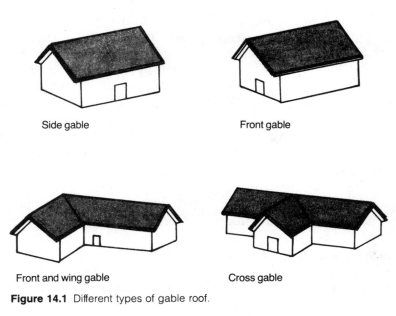

Side gable   Front gable

Front and wing gable   Cross gable

Figure 14.1 Different types of gable roof.

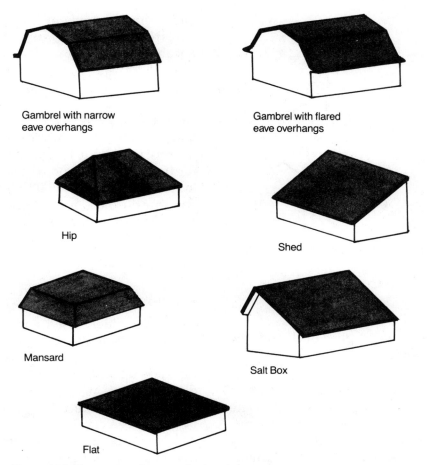

Gambrel with narrow
eave overhangs

Gambrel with flared
eave overhangs

Hip

Shed

Mansard

Salt Box

Flat

Figure 14.2 The most easily recognized roof shapes.

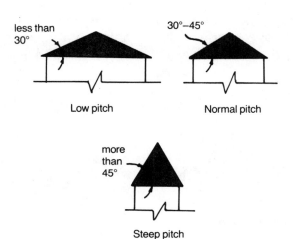

less than
30°

30°–45°

Low pitch

Normal pitch

more
than
45°

Steep pitch

Figure 14.3 Classification of roof pitches (slopes).

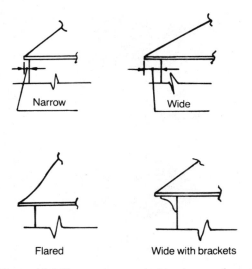

Narrow   Wide

Flared   Wide with brackets

Figure 14.4 The most recognizable shapes of eave overhangs.

a feature of the Split-Level and the Georgian styles.

Eave Overhangs The shape and size of the eave overhangs help to delineate the style of the house. The most recognizable shapes of eave overhangs are the **narrow, wide, flared,** and **wide with brackets** (Figure 14.4). The narrow eave overhang is identified with the Cape Cods, and some of the Dutch Colonials; the wide eave overhang is found in the Prairie style and some Ranches; the flared eave overhang is mostly identified with the Dutch Co-

lonial; and the wide eave overhang with brackets shows up in the Mediterranean and the Spanish and Italian Villas.

Dormers Dormers are substructures that are built into the sloped sides of the roof to provide light and ventilation for the attic or the rooms that are built in it. They can be broadly identified as window dormers or shed dormers (Fig. 14.5). Window dormers may have gable, shed, or hip type roofs.

Window Styles

The most widely used window styles are shown in Figure 14.6. The **double-hung** consists of a pair of sashes that can slide up and down. **Casement** windows consist of one or more sashes hinged at the side to open outwardly by means of a crank. **Sliding** or **gliding** windows are made with two sashes that slide horizontally, similar to sliding doors. The **picture** or **fixed** window has no moving panels. **Awning** windows open out and up by means of a crank. **Basement** windows open in and up. **Bay** and **bow** windows project outward to provide more light. The bay is shaped like a trapezoid when

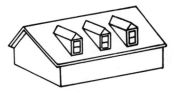

Window dormer
with gable roof

Shed dormer

Figure 14.5 Shapes of dormers.

Double-hung

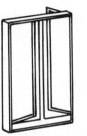

Casement
(looking from inside)

Picture or Fixed

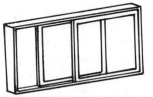

Sliding or Gliding

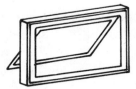

Basement
(looking from outside)

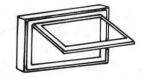

Awning
(looking from outside)

Bay

Bow

Gothic

Palladian

Figure 14.6 Window styles.

you see it from the top. The angle between the wall and the sides of the window is either 30° or 45°. The bow window is shaped more like a curve when you look at it from the top. The **palladian** window is made up of three tall windows set side by side with a half-circle window on top of the central one. The **gothic** window is long with a pointed top.

The style of the windows should match the architectural style of the house. For example, the double-hung windows go with the Colonial style houses; two side-by-side double-hung windows topped with arched molding match the French Provincials; the casement and awning windows suit the Contemporary, Ranch, Split-Level, and Mediterranean styles; and palladian windows are correct for Victorian, Shingle, and Colonials. Gothic windows are characteristic of the Gothic Revival style. Two side-by-side long windows with circular top is an identifying feature of the Mediterranean style.

Main Entrances

The main entrance includes the door and the lights around it, the columns or pilasters along its sides, and the pediment above it.

Door Styles   The door may be **paneled, flush** or **French.** Any style may have a combination of **transom lights, side lights,** and **round or elliptical top fan light** (Figure 14.7). Paneled doors are suited to the Colonials. The paneled door with elliptical fan light matches the French styles. French doors match the Mediterranean, Italian and Spanish Villas. The paneled doors with transom

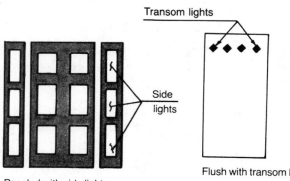

Paneled with side lights

Flush with transom lights

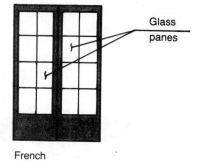

French

Figure 14.7 Door styles.

and side lights are appropriate to the Greek Revival. Flush doors are used with modern styles such as the Contemporary, Ranch, and Split-Level.

Pilasters The shape and size of the pilasters or columns can be an identifying feature of some house styles. For example, massive two-story high columns are a distinctive feature of the Greek Revival. Slimmer two-story high columns are a feature of the Southern Colonials. One-story high columns are found on some Georgians and Italian Villas.

Pediment The pediments are above-entrance decorations. Figure 14.8 shows some of the most popular pediments that are characteristic of the Georgian style. The one shown in Figure 14.9 is an identifying feature of the Greek Revival style.

Exterior Siding and Roofing Materials

The vast majority of houses are built of wood framing covered with various types of exterior siding and roofing materials designed to protect them from the weather, and to add to their beauty.

Exterior Siding Materials Exterior siding materials may be **wood shingles, wood shakes, vinyl, plywood, clapboard, aluminum, brick veneer, stone veneer, stucco, brick-face stucco,** or **half-timbering.** The building materials of the exterior siding can serve to identify the style of the house. For example, half-timbering is typical of the Elizabethan style. It is easily recognized by the exposed dark painted or stained timbering that makes up the frame structure of the walls, with the spaces between the timbers filled with bricks or stones and mostly finished with stucco. Vertical or diagonal wood siding is mostly found in the Contemporary or Shed styles, and the brick or stone veneer is used for the Colonials and Greek Revival.

Roofing Materials Roofing materials may be **asphalt shingles, wood shingles, wood shakes, slate** or **red clay tiles.** The most distinctive type of roofing material is the clay tile, which is an identifying feature of the Spanish, Italian, and Mediterranean styles.

Decorative Details

Decorative details are used to beautify the exterior of the house. They do not serve any structural function. Such details may be employed around the main entrance doors as discussed earlier, around the windows, in the arrangement and coloring of the exterior siding and roofing materials, and at the eave overhangs.

Some decorative details serve to identify particular house styles. The truss-shaped stick decoration at the apex of the gable is one of the features of the Stick style. Other types of gable

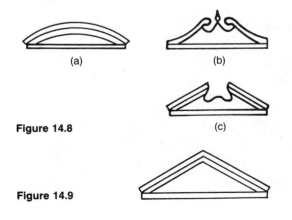

(a)

(b)

Figure 14.8 (c)

Figure 14.9

decorations are featured in the Queen Ann Victorian, and Gothic Revival styles. The tooth-shaped moldings along the cornice of the roof and portico are typical of the Georgian style.

House Styles

American house styles are diversified. Immigrants from England, the Netherlands, France, and Spain brought the architectural designs of their mother countries with them. The dominant style in each geographical region of the U.S. depended on the national origin of its settlers. The Salt Boxes and Cape Cods flourished in the New England states. The Dutch, who settled in the Mid-Atlantic states, introduced the Dutch Colonials. The Georgian and Early American (Federalist) Colonial styles dominated the Southern states. French architecture influenced the Gulf states while the Spanish styles turned up in the Pacific and Southwestern states.

With the passage of time, every style found its way all over the country. Moreover, homeowners revised them to suit their own taste. In the process, they often mixed the characteristics of two or more styles. This phenomenon is very evident in houses that are being built today, where a Contemporary look is often added to many of the old styles. These styles are renamed to match their new look: the Contemporary Tudor, the Neo-Victorian, the Contemporary Salt Box, etc.

To help the reader identify the different styles and choose the style he or she likes most, 34 of the most popular styles in varying sizes are presented in the following pages. The basic characteristics of each style and its historical origin are briefly explained. For

Figure 14.10 The Contemporary.

illustration a photograph of each is included. Let's start with the modern styles: the Contemporary, Shed, Ranch, Raised Ranch, Split-Level, and the A-Frame.

Contemporary The Contemporary style does not follow any particular pattern. It is recognizable by its "modern" asymmetrical exterior, and a roof that is a combination of gable, cross gable, side gable, and shed. Almost every contemporary roof has one or more skylights. The height and shape of the exterior walls are not uniform, and may include some curves as shown in Figure 14.10. The exterior siding is usually cedar, plywood, or any other type of wood with diagonal or vertical grooves. The roofing is usually shingles. The windows are of different sizes, shapes, and styles. The entrance door is usually recessed.

The living room, family room, the dining room, and the kitchen are most often found on the main floor. The living room usually has high vaulted or beamed ceiling. The bedrooms could be on the main floor, or some on the main floor and some on the second floor (the master bedroom would be on the second floor), or all on the second floor. Some of the Contemporary's interior features are combined rooms such as living and dining room, or kitchen and dining room; sunken family room; lots of glass and mirrors, particularly in the bathrooms. The Contemporary style has become very popular.

Shed The Shed is similar to the Contemporary. Its main feature is the multi-direction shed-type roof, as shown in Figure 14.11, that may or may not be connected with a gable roof. Its exterior

Figure 14.11 The Shed.

siding, roofing materials, windows and entrance door are very close to those of a Contemporary. In fact, most people call it a Contemporary.

Ranch   The Ranch style is sometimes called the **California Ranch** because it started in California in the mid-1930s. From there, its popularity spread to the entire country.

The Ranch is a one-story house as shown in Figure 14.12. The living, family and dining rooms, the kitchen, bedrooms, and bathrooms are all located on the main floor, which may be a few steps higher than the grade elevation.

The front of the Ranch is asymmetrical. The floor plan may be rectangle, L or U shaped. The roof may be side gable, front and wing gable, cross gable or low pitch hip. The exterior siding may be cedar, clapboard, aluminum, vinyl, shingles, brick veneer or stone veneer. The roofing is asphalt shingle for most houses and slate for the more expensive ones. The windows can be a combination of bow, bay, picture, casement, and double-hung. It usually has sliding glass doors leading to the back yard.

The Ranch style is suitable for small houses. It is also convenient for the elderly and infirm who cannot climb stairs.

Raised Ranch   The Raised Ranch has many other names: **Hi-Ranch, Bi-Level,**

Figure 14.12 The Ranch.

Split Ranch, Split Foyer or **Split Entry** The main floor is a half story above the grade and the basement and garage are a half story below the grade as shown in Figure 14.13. This arrangement allows bigger windows for the basement, which makes it usable as living space. The main entrance door opens to a staircase platform that leads to the stairs going up to the main floor and those going down to the basement.

The floor plan is typically rectangular. The roof is side gable. The exterior siding can be shingles, shakes, aluminum, bricks, vinyl or any other material. The roofing may be shingles or shakes. The windows may be double-hung, casement, or picture. The Raised

Ranch is an economical design because of the additional use of the basement.

Split-Level The Split-Level is distinguished by its front and wing gable roof in which the part of the roof covering the bedrooms is one-half story above the other part covering the rest of the house as shown in Figure 14.14. The exterior siding may be clapboard, cedar, shingles, aluminum, brick veneer, stone veneer, or vinyl. The roofing is shingles. It may have a one- or two-car garage. The windows may be double-hung, casement, bow, bay, or picture.

Some Split-Levels have the garage, the family room and the basement

Figure 14.13 The Raised Ranch.

Figure 14.14 The Split-Level.

located below the bedrooms. Others have the family room below the bedrooms and the garage and the basement below the living and dining rooms. Because the living and family rooms and the kitchen are accessed from different levels, this style is suitable for sloping grades. It has numerous floor arrangements—front to back, back to front, and side to side—to suit different grade slopes.

The Split-Level is suitable for medium-size houses. Its design provides privacy for the bedrooms, but also necessitates climbing stairs.

A-Frame   The A-Frame got its name because its front elevation or facade resembles the letter "A." The house may include either one and a half stories or two and a half stories. The roof is a steep gable that extends to the ground for the one and a half story version or ends at the top of the first floor for the two and a half story house. The main entance door is on the side. The one and a half story model is easy to construct because the sides of the roof function as exterior walls. However, this advantage is offset by the low interior ceiling resulting from the sloped roof.

The A-Frame is a small-size house. This makes it popular as a vacation or second home. But it can also be used as a primary residence. When designed for recreational purposes, the front facade is mostly glass windows, and sliding doors overlooking a deck (Figure 14.15). The front eave overhangs are wide to shade the windows from the sun. The exterior is surrounded by huge decks and patios for outdoor living. The living and dining areas are combined and located at the

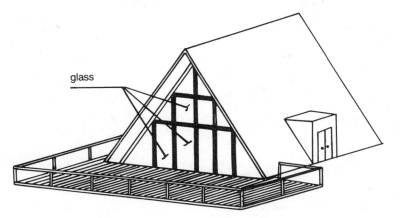

Figure 14.15 The A-Frame used as a vacation home.

front of the main floor where the big glass windows provide an unobstructed view. The kitchen is typically located in the middle and one bedroom at the rear. The rest of the bedrooms are located on the upper floor.

When the A-Frame is used as a primary residence, it usually has two and a half stories (Figure 14.16). The siding may be shingles, clapboard, vinyl, or aluminum. The roofing is shingles. The windows may be double-hung or case-

Figure 14.16 The A-Frame used as a primary residence.

Figure 14.17 The Bungalow.

ment. The exterior door may be paneled with side lights.

Bungalow The Bungalow is a native American small-size house. It was developed in 1905 and by 1920 had become very popular. One of its main features is the big porch that occupies the entire front of the house as shown in Figure 14.17. Another is its asymmetrical side gable roof in which the slope of its longer front side flattens over the porch. The original Bungalow was a one-story structure. Later, a few bedrooms were added on the second floor. Light and ventilation were provided to these rooms by a shed dormer. The roof often has wide eave overhangs to shade the windows and the porch from the sun.

The exterior siding may be clapboard, aluminum, vinyl, or any other material. The roof is usually shingles. The windows may be double-hung or casement. The floor plan of the Bungalow does not include hallways. The entrance leads to the living room or parlor, thence to the dining room, kitchen, and the bedroom(s), going from one room to another. The second floor has additional bedrooms.

Cape Cod and Cape Ann The Cape Cod is a small house that began in New England more than 200 years ago. It is classified as a Colonial. It may be

Figure 14.18 The Cape Cod.

one, one and a half or two stories high. Its roof is steep side gable (Figure 14.18), and the eave overhangs are narrow. The exterior siding may be clapboard, aluminum, shingles, vinyl, or brick veneer. The roofing is shingles. Except for the garage, the house is symmetrical. The windows are double-hung with panes, with or without shutters. Traditionally, the chimney is located in the middle of the house. However, many Cape Cods are now being built with the chimney on the side, or even without a chimney. The interior arrangements depend on how big the house is. The smallest, which is sometimes called **Cape Cottage**, has the kitchen, the living and dining areas,

and two bedrooms on the first floor. Cape Cods are becoming bigger with two or more bedrooms being added in the attic.

The Cape Ann style is similar to the Cape Cod except that it has a gambrel roof as shown in Figure 14.19.

Salt Box The Salt Box is of English origin. It is recognized by its roof, which is side gable with one side long and extending down to the first floor and the other side short and covering the second floor (Figure 14.20). The one-story part of the house is called the **lean-to** or **added-to** part.

The Salt Box got its name because

Figure 14.19 The Cape Ann.

Figure 14.20 The traditional 1½-story Salt Box.

Figure 14.21 The modern two-story Salt Box.

its exterior resembles that of the box in which salt was kept in the old country stores. It started in the New England states more than 300 years ago, and soon became popular because its design was appropriate for cold weather. The house was oriented in such a way that the long side of the roof, which had no windows, faced north to deflect the cold winter winds, while the opposite side, which had most of the windows, faced south to capture the warmth of the sun. It had a massive central fireplace to provide heating for the entire house. The oven was located next to the fireplace. The kitchen, living and dining areas were on the first floor, and the bedrooms on the second floor.

In the modern Salt Box, a second story is often added on the top of the lean-to part of the house in order to increase the number of bedrooms. Dormers are installed to provide light and ventilation for the added part as shown in Figure 14.21. The modern version of the house may be oriented in any direction, and the chimney need not be in the middle of the house, thanks to central heating. The windows are usually double-hung with panes, with or without shutters. The main entrance door may be flush with transom lights or any other type of design. The exterior siding may be shingles, clapboard or aluminum. The roof is usually shingles. The eave overhangs may be narrow or wide.

Southern Cottage   The Southern Cottage is a small native American house. It was developed by the early French colonists. Its main features are the broken side gable roof, and the entrance porch that dominates the entire front (Figure 14.22). The roof may have shed dormer(s) to accommodate more bedrooms in the attic. The eave overhangs are wide. The windows may be double-hung, bow, bay, or casement. The exterior siding may be vinyl, shingles, clapboard or any other material. The roofing is shingles.

The living room, dining room, kitchen, and either a family room or a bed-

room are on the first floor and the rest of the bedrooms are on the second floor. The house is compact, which makes it appealing to first-time homeowners and small families, and useful as a second or vacation home.

Dutch Colonial   The Dutch Colonial has its origin in the Netherlands. The Dutch settlers started building a primitive version of this style in the seventeenth century in the Dutch country of Pennsylvania and New York. (It was the Dutch who bought the Island of Manhattan from the Indians in 1624 for $24 worth of beads.)

This style can be easily identified by its gambrel roof. The eave overhangs may be narrow (Figure 14.23) or flared (Figure 14.24). The shape of the roof

Figure 14.22 The Southern Cottage.

Figure 14.23 A Dutch Colonial with narrow eave overhang.

Figure 14.24 A Dutch Colonial with flared eave overhang.

allows for a spacious attic. In fact, most Dutch Colonials have a second story in the attic. Light and ventilation for the attic rooms are provided by window or shed dormers. Many existing examples have either one or two porches on the sides of the house. These days, most Dutch Colonials are being built with a one- or two-car garage on the side.

Except for the porch and garage, the house is symmetrical. The windows are double-hung, with or without shutters. The exterior siding may be shingles, clapboard, aluminum, vinyl, bricks or any other material. The roofing is usually shingles. The entrance may include a pediment, pilasters or both.

The entrance door may be paneled with side lights, or in any other style.

Most Dutch Colonials are medium- to large-size houses. They may be one, one and a half, two, or two and a half stories. The living areas, kitchen, and dining room are located on the first floor, and the bedrooms on the second and third floors.

Georgian The Georgian style is a symmetrical two-story Colonial. It started in the English colonies at the beginning of the eighteenth century and became quite popular. It can be identified by its central main entrance door, usually capped with a pediment

Figure 14.25 The Georgian.

Figure 14.26 The Greek Revival.

that is supported either as a cantilever from the wall or by decorative columns (Figure 14.25). On each side of the door there are two windows. The second floor has five windows centered above the main entrance door and the four windows of the first floor. The roof is mostly a side gable with normal slope. Some examples have hip roofs. The edges of the cornice and the pediment typically have tooth-shaped decorative moldings. The siding is mostly brick veneer. The windows are double-hung with shutters. The roofing may be shingles or slate. The door is mostly paneled with or without side lights.

The entrance door leads to a foyer or a hall that has the living room on one side and the dining room on the other. The family room, kitchen, and laundry room are located at the rear. All the bedrooms are located on the second floor.

Greek Revival   The Greek Revival style can be recognized by its dominant central front entry porch with a huge pediment and Greek columns as shown in Figure 14.26. The gable of the pediment has two or more trim layers. The columns have capitals and bases. This style was very popular during the first half of the nineteenth century, and is gaining in popularity now.

The house is typically two stories high with low pitch side gable roof. The exterior siding is usually brick, sometimes painted white, and the roofing may be shingles or slate. The main entrance door is paneled with or without transom or side lights, and with various types of decorations around it. The windows may be double-hung with shutters, or casement.

Greek Revival houses are among the biggest. The floor plan arrangements depend on the size of the house and whether it has one or two wings. Generally one wing encompasses the garage and the mudroom. The second wing may encompass the master suite, a study room, or a library. The living and dining rooms and the kitchen occupy the first floor of the central block. The bedrooms occupy the second floor.

Early American   The Early American style is a two-story early Colonial house like the Georgian, Cape Cod, Salt Box, or Dutch Colonial. It is attached to a separate front gabled garage by means of a one story **connector** that has a side gable roof (Figure 14.27). The exterior siding may be cedar, clapboard, shingles, vinyl, brick or stone veneer. The roof is usually shingles. The windows are mostly—but not always—double-hung with shutters. The main entrance door is typically double paneled and topped by a pediment.

The floor arrangement is basically the same as in the other colonials except for the area adjacent to and including the connector. The connector

Figure 14.27 The Early American.

Figure 14.28 The French Colonial.

may include a laundry room or mud-room in the front and a nook or break-fast room in the back, the family room, or a study room. The attic of the garage may be accessed by a folding ladder so that it can be used for storage.

French Colonial Many of today's French Colonials are two-story sym-metrical structures with steep pitch hip roofs. The main entrance door is re-cessed and is usually double with el-liptical top window. The top of the entrance is either elliptically shaped or has elliptical decorations. The front of the first floor has two big windows, one on each side of the main entrance.

The front of the second floor has three windows centered above the main en-trance and the first floor windows. The siding is either white stucco or brick veneer painted white. The roofing may be shingles or slate. The windows may be double-hung, casement, or bay. The living, family, and dining rooms and the kitchen are on the first floor and the bedrooms on the second floor.

Large versions have two symmetri-cal **wings,** one on each side of the **central building (the block)**. The front line of both wings is a few feet forward of the front line of the block. Both wings are one story high and have a steep pitch hip roof (Figure 14.28). Typically, one wing includes the garage and a mudroom and the other a study room, library, the master bedroom, or a guest room.

French Provincial The French Provincial, like the French Colonial, is a big and elegant style that has its origin in France. It consists of a symmetrical two-story structure called the central block, flanked on each side by a one-story wing. The front line of the wings is a few feet forward of the front line of the block. Each section has a dominant steep hip roof as shown in Figure 14.29. The top of the second floor windows of the central block is curved and breaks through the eaves and the lower part of the roof. The windows are usually double-hung, with shutters. The top of the shutters is curved. The exterior siding is usually brick, and the roofing may be shingles or slate. The door is recessed and may or may not have an elliptical fan window at its top. The top of the entrance is elliptical in shape, or has elliptical decorations.

In a big house like the one shown in the picture, the master suite is likely to occupy one of the wings and includes the master bedroom, sitting room with a fireplace, one or two baths, two dressing rooms, and two walk-in closets. The other wing is likely to include a three-car garage, a mudroom, and possibly a breakfast nook. The central

Figure 14.29 The French Provincial.

Figure 14.30 The French Mansard or Second Empire.

block includes the living room, dining room, family room and the kitchen on the first floor, and the bedrooms on the second floor. Each bedroom may have a separate bathroom and walk-in closet.

French Mansard or Second Empire
The French Mansard was named for François Mansart, the seventeenth-century French architect who often employed a distinctive Italian-derived roof line in his work. This style was very popular in France during the Second Empire of Napoleon III (1852–1870). It was widely used in the U.S. during the late nineteenth century. The ex-

amples that were built during that time have elaborate roof and dormer decorations. The roof comes in three styles: concave, convex, and straight.

Most of the Mansards that are being built now have simple roof lines as shown in Figure 14.30, for economy's sake. The windows are usually double-hung with top curved moldings, and shutters with curved tops. The main entrance door is recessed. The exterior siding may be wood, vinyl, or brick veneer and the roofing material is shingles. Except for the garage, the house is symmetrical. The garage may or may not have a second story.

The first floor arrangement is similar

to the colonials except that the family room may be located at the front and the dining room at the back next to the kitchen. As in all two-story houses, the bedrooms are on the second floor.

Italian Villa   The Italian Villa was developed in the United States during the middle of the nineteenth century. It can be identified by its flat or very low pitched hip roof, its wide eave overhangs with brackets, and its tall and narrow windows (Figure 14.31). The front may or may not be symmetrical. It has a dominant front porch the roof of which has wide overhangs with brackets. The outer edge of the porch is supported by decorative columns. The windows may have top decorations. The siding may be cedar, clapboard, vinyl, stucco, or brick veneer. The roofing has to be built-up asphalt, because of its low pitch. The main entrance door is likely to be French with decorative pediment at its top and pilasters along its sides. The windows may be double-hung with shutters, or casement.

The floor arrangement depends on the size of the house. Generally, the first floor includes the living room and dining room in the front, the family room, kitchen, and laundry room in

Figure 14.31 The Italian Villa.

Figure 14.32 The Mediterranean.

the back. The bedrooms are on the second and third floors.

Mediterranean The Mediterranean style is a two-story symmetrical structure. It can be identified by its low pitch hip roof, clay tile roofing, wide eave overhangs with brackets, large first floor windows with circular top and ornamental railings at the bottom (Figure 14.32). The main entrance is projected outward. The projection may stop at the first floor or extend to the second floor. The entrance door may be French with elaborate decorations around it. The second floor windows are smaller than those of the first floor and may be double-hung, or casement. The exterior siding may be stucco or painted brick veneer.

The first floor arrangements depend on the size of the house. Usually a dining room is on one side of the front and a living room on the other side. The family room, kitchen, and mudroom are in the back. A big house may include a library or a guest room. The bedrooms are found on the second floor.

Spanish Villa The Spanish Villa may be one or two stories high. Its front may or may not be symmetrical. It can be identified by its low pitch hip roof with wide eave overhangs and by the dominant front porch enclosed by big arched openings (Figure 14.33). The arched openings have ornamental iron railings. The part of the house overlooking the porch may have balconies with ornamental iron handrailings that match those of the arched openings. The windows may be casement or double-hung, with or without arched tops. The main entrance door is paneled with a half-circle window on top of it. The balcony doors are likely to be French. Generally, the exterior siding is stucco and the roofing is clay tile.

The floor arrangement depends on the size of the house. The first floor is likely to include the living room, the kitchen, the dining room, the family room, the mudroom, the garage, and possibly a study room or a guest room. The bedrooms are on the second floor. The example shown in the picture is likely to have a two-story entrance hall. This is evident by the height of the entrance door and the absence of a window above it.

Mission   The Mission style is native to America. It started in California in the late nineteenth century, and spread from there to the rest of the country. Its most obvious feature is the tower with pyramid shaped roof resembling the bell tower of the old mission churches, from which it got its name

Figure 14.33 The Spanish Villa.

Figure 14.34 The Mission.

(Figure 14.34). A porch is also a part of this style (located next to the tower in the photograph). Here we see many Spanish features: the red clay roof tiles, the wide eave overhangs with brackets, the arched main entrance, and the stucco siding.

There is no standard arrangement for the Mission style. Generally, the first floor includes the living room, dining room, kitchen, and garage. The bedrooms, and possibly a sitting room, are found on the second floor, and the third floor may have another bedroom or two.

English Tudor The English Tudor style became popular in the U.S. during the 1920s. Some of the houses that were built during that period are massive with huge chimneys (Figure 14.35). The roof is steep side gable with one or more dominant steep cross gables, and several dormers. The exterior siding may be brick or stone. The roofing for a house like the one shown in the picture must be slate. The windows are casement or picture with big panes. The main entrance door is typically paneled and surrounded by various types of decorations.

The floor arrrangements vary from house to house. The first floor generally includes the living room, parlor, sitting room, library, a huge kitchen, dining room, and mudroom. The sec-

Figure 14.35 The English Tudor.

ond floor features five or six bedrooms. The rooms tend to be large. The ceilings of the living and family rooms are usually vaulted with dark stained oak beams. The top portion of the interior walls is stucco and the lower portion is wood panel. Interior decorations can be elaborate and include lead glass casement windows, dark oak doors, and wrought iron railings for the stairways.

Elizabethan Half-Timber and Neo-Tudor The Elizabethan Half-Timber style is one of the Tudor substyles that had its origin in England during the sixteenth and seventeenth centuries. Most of the Elizabethan Half-Timbers that we see today were constructed during 1920s. They vary in size from medium to very large. Many of them are quite expensive.

This style can be easily recognized by the exposed heavy timbering that constitutes a part of the supporting structure. The spaces between the timbering are filled with brick or stonework, usually finished with white stucco.

Today's Half-Timbers are constructed of regular wood framing with stucco siding. The exposed heavy timbering is replaced by dark colored stucco as shown in Figure 14.36 for economy's sake. Many examples have a contemporary look, and thus are referred to as **Neo-Tudor** or Tudor Adaptation.

The roof of the Elizabethan style is steep pitched side gable with one or more prominent cross steep gables. The roofing material is usually shingles or slate. The windows are either casement or double-hung with diamond

Figure 14.36 Today's Elizabethan Half-Timber or Neo-Tudor.

shaped panes. The entrance door may be all solid, or half solid and half glass with diamond panes. The interior arrangement may be Colonial, Ranch, Split, or any other style—with Elizabethan Half-Timber detail.

Cotswold Cottage   The Cotswold Cottage is a small Tudor. Its front is asymmetrical and is dominated by a big steep gable with a huge projected stone or brick chimney located at the gable's center (Figure 14.37). The main entrance is located at the side of the big gable and is topped by a small gable with a pitch identical to that of the big gable. The siding is mostly white or gray stucco. The roofing is shingles or slate. The windows are double-hung or casement with rectangular or diamond shaped panes.

The main entrance door opens to a sitting room with a beamed ceiling. This room functions as both a formal living room and a family room. The kitchen and dining room are next to the sitting room. The first floor may also include one or two bedrooms set a few steps higher than the main floor. The rest of the bedrooms are on the second floor. Usually, the upper part of the interior walls is stucco and the lower part is paneled in wood.

Norman Tudor   The Norman Tudor style is suited to a medium-size house. It can be identified by its roof, which is mainly a side and wing normal-to steep-gable (Figure 14.38). The main entrance door is located on the side of the wing gable and has a pediment that matches and is parallel to the roof lines. The exterior siding is mostly white stucco with half-timber imita-

Figure 14.37 The Cotswold Cottage.

Figure 14.38 The Norman Tudor.

Figure 14.39 The Vernacular Wright Prairie.

tion. The roofing may be shingles or slate. The windows may be double-hung, casement, or bay with square or diamond shaped panes. The main entrance door is paneled, with or without top arched window. Most examples have a massive chimney on the side.

The main entrance door opens to a sitting room, which acts as both living and family room. It may have a cathedral or beamed ceiling. The kitchen and dining room are located in the back. A few steps lead up to the bedrooms. The second floor may include another two bedrooms. The upper part of the interior walls is stucco and the lower part is paneled in wood.

Prairie The Prairie house was developed by American architect Frank Lloyd Wright at the turn of the twentieth century. This style comes in many seemingly unrelated varieties.

The most common substyle is the one shown in Figure 14.39. It is sometimes called **Vernacular Wright.** Its plan is square or rectangular and its roof is a hip with dormers. The front view is symmetrical, usually with a big porch. The siding may be clapboard, aluminum, shingles, or any other material and the roofing is shingles. The windows are usually double-hung. The main entrance door is simple and unelaborate. The floor plan may be arranged so that the first floor is independent from the upper floor, which makes it convenient as a mother–daughter, in-law, or two-family accommodation.

Figure 14.40 The Organic Architecture Prairie.

Another Prairie substyle is shown in Figure 14.40. Wright called it **Organic Architecture.** It can be identified by its flat or low pitch roof with wide eave overhangs, massive central chimney, and short, wide windows. The windows of the second floor are continuous so as to provide an unobstructed view. The back of the house is predominantly glass with French doors.

Queen Ann Victorian   The Queen Ann Victorian is an asymmetrical structure with many variations or substyles. It was popular during the late nineteenth century, and is gaining popularity again. Most Victorian houses are two and a half or three stories high. The roof is irregular with gables facing several directions. The first floor entrance has a dominant porch, usually L shaped (Figure 14.41). The second floor may have a smaller porch. In many examples, a big part of the front projects outward in a large broken arc. The exterior siding may be shingles, shakes, vinyl, or any other material. The roofing is shingles. The windows are mostly double-hung. The main entrance door may be French or any other type.

The plan of this style allows each floor to be independent. The main entrance door opens to the stairs leading to the upper floors. The living room is usually located behind the dominant curved front. The family room, kitchen,

Figure 14.41 The Queen Ann Victorian.

and bedrooms are located in the middle and rear parts of the house. Usually, there is another small staircase at the rear of the house, next to the kitchen.

Gothic Revival   Gothic Revival is of English origin. It reached the peak of its popularity during the mid-nineteenth century. Its most obvious feature is its very steep multi-cross gabled roof (Figure 14.42). The eave overhangs are wide and are often elaborately decorated. The dormers have the same pitch and overhang decorations as the main roof. The central second floor window may be in Gothic style, but this is not always the case. The front may or may not have a porch. The siding may be clapboard, vinyl, aluminum, stucco, or brick veneer. The roofing may be shingles or slate. Usually, the windows are double-hung, and the main entrance door has glass side lights.

The first floor arrangements depend on the size and shape of the house. Here you will usually find the living room, dining room, family room, kitchen, and laundry. The bedrooms are on the second and third floors.

Stick   The Stick style is a simple version of Gothic Revival. It also bears some resemblance to Queen Ann Vic-

Figure 14.42 The Gothic Revival.

torian. It was popular during the 1870s and 1880s. Its roof is front steep gable, frequently with secondary cross steep gable. The eave overhangs are wide. Its main identifying feature is the decorative stick trusses located at the apex of the gables (Figure 14.43). In many cases, horizontal narrow banded siding marks the floor's elevations, and vertical narrow banded siding is found along the interior walls or partitions. The exterior siding may be clapboard or shingles, and the roofing is typically shingles. The main entrance has a prominent porch of rectangular or L shape. The top of the columns supporting the porch have diagonal decorations. The windows may be double-hung or bow.

Most Stick houses are narrow and deep. The plan can be such that the first floor is independent from the upper floors, which allows it to accommodate a mother–daughter, in-law, or two-family arrangement.

Shingle The Shingle style started in the U.S. in the late nineteenth century. It can be identified by shingle walls that are continuous at the corners. Some examples have **turrets** as shown in Figure 14.44. The front is asymmetrical with a dominant first floor porch. Both the floor plans and the roof lines are irregularly shaped. The roof is steep pitch multi-cross gables, mostly done in shingles. The windows may be dou-

Figure 14.43 The Stick.

Figure 14.44 The Shingle.

ble-hung with or without shutters, palladian, or bay. The main entrance door may be paneled or any other style. Large examples may have two main entrances, one at the front and the other at the side.

The floor arrangement varies depending on the shape and size of the house. Generally, the first floor includes the living room, family room, dining room and kitchen, and the bedrooms are found on the second and third floors. Some examples have two bedrooms on the first floor, which makes it independent of the top floors. This allows the house to be occupied by two families.

Farmhouse The Farmhouse can be identified by its porch with shed-type roof. It extends over a large portion of the front, its outer edge supported by curved beams that are in turn supported by square columns (Figure 14.45). (The beams and columns are usually painted white). The house may be one, one and a half, or two-stories high. The one-story house is sometimes called **Traditional**. The roof is generally side gable, but some examples have front and wing gables. The exterior siding may be shingles, clapboard, vinyl, cedar, aluminum, or brick veneer. The roofing is mostly shingles. The windows are mostly double-hung, and the exterior door has transom and side lights. Most Farmhouses built today have some Ranch, Colonial or Contemporary characteristics. However, all of them include the shed-roof porch.

Figure 14.45 The Farmhouse.

Figure 14.46 The Townhouse.

The arrangement of the one-story Farmhouse is similar to that of a Ranch. That is, the family room, living room, dining room and kitchen are in the middle. The garage, if any, is at one end, and the bedrooms are at the other end. The arrangement of the one and a half or two-story Farmhouses feature the living room, family room, dining room and kitchen (and possibly one bedroom) on the main floor and the rest of the bedrooms on the second floor.

Townhouse The Townhouse is a narrow and deep rectangular structure that may be two, three, or four stories high. Some Townhouses are detached (Figure 14.46), but most are semi-attached or completely attached to adjoining houses by means of **party walls.** Townhouses are common in cities where land is scarce and expensive because they do not require much **front footage.**

A Townhouse may be in any style: Italian, French Mansard, Colonial,

Greek Revival, etc. Many of the examples being built today are three stories with a flat or Mansard roof. They are arranged so that the homeowner may occupy the first and second floors, and rent out the third floor. The first story includes a garage, a family room, and a storage area. The second story features two bedrooms, a kitchen, a bathroom, and a spacious living room that is used also for dining. The third floor arrangement is similar to that of the second floor, with an independent entrance.

CHAPTER 15

Choosing a Design

*C*hoosing a good design for your house requires planning. Start with listing all your needs and desires such as a kitchen with a nook or breakfast room, or a large living room to entertain guests, or a workshop in the garage or the basement. Each member of the family should have his or her input. Put all the ideas on pieces of paper, even though they may change later. It is recommended that you file all these papers in a three-ring binder so that they do not get lost.

Meanwhile, increase your knowledge about various aspects of home design.

1. Sift through home plan (stock plan) books, and architectural magazines and books.

2. Tour the neighborhood and note the houses you like. Write down what you like and why.

3. Stop by houses under construction and chat with the workers.

4. Visit lumber yards and showrooms, particularly those for bathroom fixtures, windows, doors, and building materials.

5. Talk with friends and neighbors who

have built houses recently to learn from their experience.

Planning

In designing your house you should concentrate on the following points: (1) the size of the house you can afford, (2) the style of the house you like the most, (3) the size, shape and topography of the lot and the effect they may have on the design, and (4) energy conservation measures that have to be incorporated in the design.

The Size of the House The size of a house is measured by the number of square feet of all the floors. You can get a feeling of house sizes by looking at various designs in home plan books and noticing the square footage of each design, and by touring houses under construction and asking about the square footage of each house. After a while, you will develop a sense of house size.

You can estimate the size of the house you can afford by dividing your budget by the estimated cost per square foot. An average estimated cost can be obtained from local general contractors and friends or relatives who have recently built their own houses.

House Styles   House styles are numerous. The preceding chapter includes 34 of the most popular ones. After studying each one, tour the neighborhood and try to identify the style of each house. The entire family can participate in this entertaining and learning experience. After some practice, picking out the different styles will become second nature.

You can also increase your knowledge about design styles by reading architectural books and magazines available in public libraries and book stores.

The Building Lot   The size, shape, and topography of the lot may impose some restrictions on the design of the house. For example, if the grade is markedly sloped, you may have to choose a Split-Level or Contemporary style. If the lot is narrow, you may have to choose a multi-story design or townhouse, since that's all the lot can accommodate.

Energy Conservation Measures   Energy conservation measures must be a part of initial planning because some of them may have to be incorporated in the design. Some of these measures are:

1. Increasing the thickness of the exterior wall to six inches to accommodate more insulation.

2. Installing a solar heating system.

3. Locating the fireplace in the middle of the house to reduce heat loss.

4. Dividing the house into two heating zones with clock thermostats for each floor. This requires installing hot-water heating, because this kind of zoning cannot be accomplished by a hot-air system.

5. Choosing windows with wood rather than aluminum sash, because wood is a better thermal insulator.

6. Installing a thermostat-controlled attic exhaust fan. It cools the house significantly during hot sunny days, which reduces the need for air conditioning.

7. Reducing the size or number of the windows facing north without upsetting the architectural balance of the exterior.

8. Locating the central air conditioning outdoor unit in a shaded location such as the north side of the house.

9. Choosing a design in which the kitchen is located next to the family/sitting room to benefit from the heat produced by the oven and the refrigerator (the refrigerator emits heat).

You should also check with the energy department of your state and the local electric utility company to see whether they have publications regarding energy conservation measures. They will be happy to send you what they have, usually free of charge.

Buying a Design

There are several ways to buy a house design: (1) from **ready-made design books** (also called **home plan books** or **stock plan books**), (2) by retaining (hiring) **an architect** or a **building designer** to custom design your house, and (3) by using a house you saw either in the neighborhood or in one of the home plan books as a base line and hiring an architect or a building designer to modify the design to suit your requirements.

Ready-Made Design Books Ready-made design books are advertised through home design and building magazines, and can be bought through mail order. There are about 20 companies in this field, each with a staff of architects and building designers who continually create new designs and update existing ones. Some books are devoted exclusively to one or a few styles. Others present designs based on the size of the house: one-story designs under 2000 square feet, one-story designs over 2000 square feet, multi-level designs, etc. Each book includes hundreds of designs to suit various tastes and fit a wide range of budgets. Each design includes an isometric of the exterior, which gives a clear picture of what the house will look like, and a plan view of each floor. The plan views are easy to read with the bedrooms, living rooms, working areas (kitchen and laundry room), and baths clearly marked and fully dimensioned. The designs are of good quality and reflect the current styles.

By itself, going through these books is a learning experience. They are worth looking at even if you intend to hire an architect or building designer, because they will broaden your horizon as far as home styling is concerned. You and your family will spend many enjoyable hours comparing designs and envisioning your dreamhouse.

Once you have decided on a particular design, you can order the corresponding detail drawings or blueprints by mail. The first set costs between $100 and $150, and each additional set costs about $20. They will be shipped to you in a matter of days. On average,

each set consists of six drawings, a bill (list) of materials, and condensed specifications. The drawings are to scale, fully detailed and dimensioned, and adequate for obtaining the building permit and constructing the house. Most designs are available also in reverse (mirrored).

Prospective homeowners buy pre-designed or stock plans for various reasons. Foremost is that they cost less than what an architect will charge for a custom design. Another reason is that you can visualize the house before you buy the design. A third reason is that professional architects or building designers are not available in many communities, particularly in rural areas.

Predesigned plans have several limitations, however. First, they are standard, and are not made to fit any particular lot. Second, they may not exactly match your needs or taste. Thus, it is very likely that you will make some changes in the plans. Some common changes are:

1. Modifying the foundations.
2. Changing the material of the exterior siding.
3. Changing the style of the windows and exterior doors.
4. Enlarging or reducing the size of one or more rooms.
5. Changing the arrangement or the style of the kitchen cabinets and bathroom fixtures.
6. Enlarging or reducing the size of the garage.
7. Omitting or adding, reducing or enlarging the basement.
8. Increasing the thickness of the exterior walls to six inches to accommodate more insulation.

9. Increasing the number of heating zones.

10. Changing the materials of the flooring.

11. Adding a cathedral or beamed ceiling to the living or family rooms.

It is important that you do not make changes that may upset the balance of the exterior of the house, such as drastically changing the size of the windows or exterior doors.

Hiring an Architect or a Building Designer

You may want to hire an architect or a building designer for a custom design. The following are situations in which the services of an architect must be sought: (1) you have specific requirements, (2) your lot is oddly shaped or has steep slopes, (3) you are planning to build a big house, say, over 4000 square feet.

Before hiring an architect, you should have the following material ready: (1) a list of all your requirements, (2) an estimate of how much money you are prepared to spend, (3) a survey of the lot, and (4) deed restrictions, if any.

Before starting the design, the architect conducts a field survey whereby he inspects the lot and the styles of the neighboring houses in order to create a design that is compatible with the surrounding environment. He creates several conceptual designs or proposals. To help you visualize these proposals, he makes sketches showing the floor plans, the exterior of the house, and possibly a three-dimensional model of each concept. He discusses these proposals with you and keeps revising them according to your suggestions until you are finally satisfied that one of them meets your requirements.

Subsequently, the architect or building designer finalizes the design drawings, writes the material and construction specifications and the bill of materials. After the building department approves the drawing, the architect gives you copies to use for construction and for your files. Usually, the architect will not give you the original tracings.

Modifying a Base Line Design

This method falls between buying a ready-made design and hiring an architect or a building designer to design your house from scratch. You start with a base line design—one you may have seen in a design book or a house in which a friend or relative lives. Next, you hire an architect, a building designer, or a draftsperson to modify this design to fit your requirements. The advantages of this method are that you know at the outset what the final design will look like, and the design fee will be less than that of a new design.

The Professionals

Most building departments require that houses be designed by a state licensed professional whose knowledge and experience have been proven. It is a requirement in many localities that the design drawings must be signed and sealed by a state **registered architect (RA)** or a **professional engineer (P.E.)**. There are other professionals such as **structural engineers, geotechnical engineers, heating engineers,** and **draftspersons** whose services may be needed.

Here is a description of the qualifications and functions of these professionals:

Architect An architect is educated and trained to transform his client's ideas, tastes, and requirements into a liveable house. The interior must be functional with easy access to each room or space and from one space to another. The exterior must be balanced, meaning pleasing to the eye. Naturally, what pleases one person may not please another. This is why it is very important that the architect understands your requirements and tastes. The architect also devotes considerable effort to the details whether they are functional or decorative.

An architect can obtain an RA license if he or she has had at least five years of architectural studies and three years of related experience and passes a state license examination. An architect may become a member of the **American Institute of Architects (AIA)** after meeting the Institute's stringent requirements. Architects are qualified to supervise house construction.

Building Designer A building designer is also educated and trained to design houses. A designer can become a member of the **American Institute of Building Designers (AIBD)** if he or she has a certain number of years of education and experience. Building designers are qualified to supervise house construction.

Structural Engineer A structural engineer is educated and trained to design the supporting system of the house and to ensure that it will last through-out the life of the structure. The supporting system includes the foundations, load-bearing walls, floor joists, and steel girders supporting the first floor above the basement or spanning the two-car garage. Most structural engineers have at least a B.S. degree, many have M.S. degrees, and a few have Ph.Ds. A structural engineer may obtain a state professional engineer (P.E.) license after meeting the state requirements and passing its examination. A structural engineer may become a member of the **American Society of Civil Engineers (ASCE)** after he or she meets the Society's stringent requirements. Structural engineers are qualified to supervise construction.

Professional Engineer A professional engineer (P.E.) is an engineer who has met the state requirements and passed its license examination. He or she may have studied structural, mechanical, electrical, or any other engineering discipline including sanitation. The P.E. examination is diversified to ensure that the licensee has well-rounded experience. Professional Engineers are qualified to supervise construction.

Geotechnical Engineer A geotechnical engineer's expertise lies in analysis of the soils, determining their bearing capacity and the most suitable type of foundations. Geotechnical engineers do not supervise house construction.

Heating Engineer A heating engineer prepares the heat sheet and sizes the radiators, baseboards, furnace and ducts. He or she does not supervise house construction.

Draftsperson A draftsperson may have one or two years of college education in drafting plus several years of experience. He or she can transform rough sketches into drawings, but does not create designs. A draftsperson may supervise construction under the direction of an engineer or architect.

Hiring Professionals

Most people are intimidated when they hear the word "architect" or "engineer." They think that these professionals must charge a fortune. Nothing is further from the truth. With the exception of a few at the top, the average hourly rate of an architect or engineer is comparable to that of a plumber, mason, or carpenter.

Why You Should Hire a Professional A house is a tremendous investment. For a lot of people, it is the greatest achievement of their lives. Any mistake in design or construction can be costly. Therefore, you must not take chances. This does not mean that you should pay unreasonable professional fees. The question is, how much is reasonable? There is no definite answer to this question. In my opinion, the fees for design and construction supervision should be within five percent of the total cost of construction unless it is a very big house or a difficult lot to work with.

How to Find Professionals There are several ways to find professionals:

1. Through relatives and friends.

2. By checking houses under construction.

3. By consulting the local chapters of the professional societies and associations:
   For architects: the American Institute of Architects.
   For building designers: the American Institute of Building Designers.
   For structural or geotechnical engineers: the American Society of Civil Engineers.

Selecting a Designer The first step in selecting an architect or a building designer is to obtain some names and telephone numbers. Call them and explain your project and ask if they are interested in designing your house. If they are, ask them about their qualifications and affiliations and how they charge their fees—on a lump sum basis, per hour, or per square foot of the area of the floors? Also ask them if they are willing to show you some of their previous work and give you some references.

When you meet with an architect or a building designer, ask to see some of his previous work. It is very important that you like his designs. This is even more important than his qualifications. Then, start discussing design costs. If it is on a lump sum basis, he should state clearly what is included (how many drawings, the bill of materials, and specifications). He should also state that he is responsible for obtaining the approval of the building department. If the cost is on an hourly basis, you should get a firm commitment on a ceiling on the number of hours, barring unforseen circumstances. The designer should give you a delivery schedule of the drawings.

If soil conditions are poor, you must consult a geotechnical or structural engineer. You may also consult the building inspector in this regard.

Selecting a Construction Manager It may be to your advantage to hire a different professional to manage construction. This will give you the benefit of another opinion. Design and construction are two different matters and a good designer may or may not be a good construction manager. Construction may be managed by an architect, structural engineer, professional engineer, or building designer. It is best that the construction manager be contracted on hourly basis (or per visit) so that you pay him only when you need his services. Interview several professionals and make your choice based on his or her qualifications and hourly rates.

CHAPTER 16

House Additions

Your need for a bigger house may stem from an addition to the family, or an increase in income that you would like to enjoy in the form of a more prestigious home. There are several situations in which making an addition to an existing house ought to be considered:

1. You like the house you are living in, or the neighborhood or the schools, and do not want to move. You just wish you had more room.

2. You have a nice, small mortgage with low interest rate and monthly payments, and you feel that you cannot afford to buy a bigger house.

3. You are in the market to buy a house and find that the one you want is beyond your means. You consider buying a smaller house you can add to later. (However, before buying such a house, be certain that it can accommodate the additions you plan to make.)

Additions may be made horizontally, meaning building a new structure adjacent to or adjoining the existing house; or vertically, meaning adding rooms on top of the existing house.

The Horizontal Addition

The size of the horizontal addition is governed by the size and shape of the lot and the local zoning ordinances. In order to determine the area in which you can build, start with a recent copy of the land survey. Next, find out the setback, right-side yard, left-side yard, and rear yard from the local building department. Mark these setbacks on the land survey in pencil. Figure 16.1 illustrates how this is done. The area between the setbacks and the existing house is where you are allowed to build the new addition. If you do not have drafting experience, you may hire a land surveyor or a draftsperson to do this sketch for you. Of course, an architect or engineer is qualified to do the job but is likely to charge you more.

The Vertical Addition

The main factors in determining the size and shape of the vertical addition are the size of the existing house, its structural strength, the ability of the foundations to support the added load, and the maximum height allowed

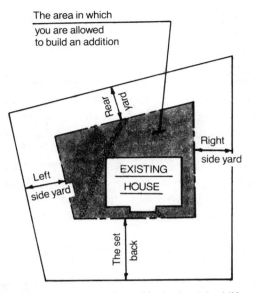

Figure 16.1 The area allowed for horizontal addition.

by the zoning ordinances. The structural elements that must be checked are the bearing soil, the foundations, the intermediate columns, the joists of the attic, and the load-bearing walls.

The Bearing Soil The bearing soil condition can be checked by inspecting the walls of the foundations and basement. The absence of diagonal cracking in these walls is an indication that the soil is good. On the other hand, if the walls do show diagonal cracking (Figure 16.2), it is a sign that the soil is exhibiting uneven settlement, and that it should not be burdened with any additional load. The ability of the bearing soil to support the additional load must be determined by a geotechnical or structural engineer.

The Foundations The foundations must be in good physical condition in order to be able to support the new load. If the walls of the foundations display horizontal or vertical cracking, or if they are leaking or their surface is deteriorating, these are all signs that the foundations may not be able to take any additional load. The ability of the foundations to support the addi-

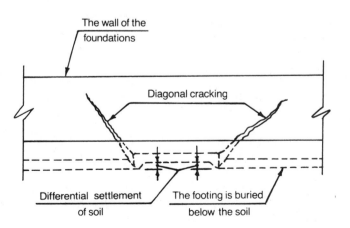

Figure 16.2 Diagonal cracking of the foundation caused by differential settlement of soil.

tional load must be determined by a structural engineer.

The Attic

The floor and roof of the attic must be checked thoroughly by a structural engineer. The attic's floor joists are usually 2 × 6 inches since they are not designed to carry any flooring. However, these joists should be increased in the areas where new floors are to be installed. The conditions of the existing rafters (the inclined joists) and the roofing should be thoroughly examined.

The Bearing Walls

The integrity of the bearing walls that will carry the additional load must be checked by a structural engineer. The parts that should be inspected are the vertical studs, and the top and sole plates. Since all these items are covered by the drywall or stucco, and cannot be seen, inspection is done by checking whether the wall surfaces are buckled or cracked. This gives a clue as to the condition of the wood framing inside.

These tips are for your guidance only. They are not a substitute for complete on-site inspection that should be carried out by a structural engineer.

The Architectural Design of the New Addition

The design of the addition is governed by the following factors:

1. Your allocated budget.

2. The shape, size, and architectural style of the existing house.

3. The size and style of the neighboring houses.

The Allocated Budget

The amount of money you are allocating to the addition determines its size. By asking local general contractors or neighbors who have recently added to their houses, you can get an idea about the average cost per square foot. Take into consideration that the cost per square foot for a bathroom is much higher than that for a bedroom or living room. By knowing your budget and the average cost per square foot, you can determine the square footage of the addition.

It is important to earmark between 10 and 20 percent of the initial budget as reserve to cover unforseen cost increases. Some increases may result from previously undiscovered deterioration of the framing, piping, or electrical wiring. Also, you may decide later on a bigger bathtub or a tub with a whirlpool, or more expensive ceramic tiles, kitchen cabinets, windows or doors.

The Style of the Existing House

The style of the new addition should be compatible with the style of the existing house. Ideally, the new addition should look as if it were a part of the original design.

1. The style of the new addition (and its roof line) should be compatible with the style of the existing house. (See Chapter 14 for help in defining the style of your home.)

2. The size, location and style of the new windows and doors should be in proportion to those of the existing house.

3. The exterior siding of the addition should match and line up with that of the existing house.

4. The connection between the new addition and the existing house should be seamless.

The interior of the new addition must be integrated with the existing house without creating bottlenecks. Also, the addition should not obstruct sunlight to the existing house.

The Size and Style of the Neighboring Houses   You should take into consideration that the size of the enlarged house should not be much bigger than the neighboring houses. It is not financially sound practice to expand your house to include five or six bedrooms if the neighboring houses have only two or three, because the resale value of your house will not appreciate in proportion to the costs of such an addition.

While you are visualizing the shape and size of your house's addition, tour the area and observe the shape, size, and architectural styles of the neighboring houses in order to make sure your added-to house will remain compatible with them. And if you know of a neighbor who has recently added to her house, talk to her about her experience. If you have hired an architect, take him with you because you both may develop some new ideas as a result of your conversation with the neighbor.

The Existing Utility Systems

The existing utility systems (heating, hot water, electricity) must be checked to see if they can support the new addition.

Heating   Your added space needs to be heated. It is doubtful that your existing boiler or furnace will have the capacity to provide the required heat. Your options are to either add a new furnace or replace the existing one with a more efficient unit. The latter alternative is attractive if the existing system is old, inefficient, and has almost exhausted its useful life. If the existing system is in good condition, or if you do not have the budget to replace it, you may consider an independent unit for the addition. You should investigate one of the compact units on the market today; a plumber can advise you on this matter.

Water Heater   The water heater is another item that should be investigated. Some older water heaters are combined with the boiler, which is efficient during the heating season and very inefficient during warm weather. If you have one of these systems, you should think seriously of replacing it with a separate, well insulated water heater. Compact water heaters are available; look at one of these before deciding on a system. Again, you should consult a plumber.

Plumbing   The existing plumbing system should also be checked to see how it will accommodate the new addition. A plumber should be able to advise you on the best location for the new bathroom fixtures and kitchen sinks.

Electrical Wiring   The existing electrical current should be checked to see if it is adequate for the new addition. The current is measured in Amperes (Amps) and it should be marked on the fuse box—for example, 100 Amps. An electrician should be consulted on this and all electrical matters.

As you can see, putting an addition on your house is not a simple matter. You would do well to consult an architect.

CHAPTER 17

Applying for a Building Permit

Before construction starts, you must apply for and obtain a **building permit** from the building department of the city or town where the property is located. It is recommended that you obtain a building permit application from the **town engineer** at the outset and learn about the zoning ordinances, state and local building codes, the energy conservation code, the documents that must be submitted with the application, and the inspection schedule. Generally, the required documents include:

1. Two sets of the design drawings.

2. A survey showing the boundaries of the lot and the position of the house within these boundaries.

3. A work sheet for heating (and cooling if central air-conditioning is to be installed) indicating the heat loss (and gain) through the walls, ceiling, floor, foundations, windows, and doors of the house. You may be required to sign a statement indicating that you will comply with the state heating code.

4. A heat sheet (not always required) showing the size and location of the air ducts, registers, returns, and the furnace for hot air heating; and the size and location of the water pipes, baseboards, radiators, and the boiler for hot water heating.

5. A certificate indicating that the lot has passed soil percolation tests if the area is not served by public sewers and a sewage disposal system is needed.

6. A certificate indicating that the water has passed the biological and chemical tests if the house is to get its water from a well.

7. A certificate that either you or the general contractor carries a Workmen's Compensation insurance policy.

Zoning Ordinances

Zoning ordinances are laws and regulations established by local governments that set standards for the use of land and the buildings to be erected, within each designated area or zone. The purpose of these regulations is to promote the health, safety, welfare, and orderly growth of the community.

These are the most common subjects of zoning ordinances:

1. The minimum area of a building lot.

2. The maximum number of stories or maximum height of a house.

3. The minimum **setback** from the street to the front building line.

4. The minimum **right- and left-side yards** from the side boundary lines to the side building lines.

5. The minimum **rear yard** between the rear boundary line and the rear building line.

6. The minimum distances from the outer edges of a driveway and walkway to the closest boundary lines.

7. The percentage ratio between the area of the house and the area of the lot.

If an owner of a lot cannot meet a particular zoning ordinance due to reasons beyond his control, he can appeal to the zoning board for a **variance,** which is an exclusion from a zoning ordinance. For example, if the lot's **topography** (shape and slopes of its surface) dictates that the house has to be built closer to the street than the zoning allows, the board may grant the owner a variance to exclude the house from the minimum setback. However, an owner cannot seek a variance just to save himself some money, for instance, building the house close to the street to reduce the length of the sewer and water lines and the driveway.

Building Codes

Building codes are technical specifications and guide lines established by state and local governments for the purpose of setting minimum construction standards and safety procedures that must be followed when building a house. This includes the type of soil that supports the foundation, the method of constructing the foundation, the method of installing the framing, the diameter and material of the water pipes, the size and safe installation of electrical wiring, the proper ventilation of the fireplace, the thickness of the drywall or sheetrock, and so forth. The building code of one locality may differ from that of the next neighborhood even if both are in the same state. For example, one town or village may accept plastic water pipes while the neighboring village accepts only copper ones.

Energy Conservation Codes

States have **energy conservation codes** that determine the minimum amount of insulation to be installed in the walls, ceilings, foundations, basements, crawl spaces, floors above unheated areas, and around hot water pipes and air ducts. These codes specify also the minimum insulating characteristics of the windows and exterior doors. In addition, each locality sets the **outside and inside design temperatures** that must be used for sizing the heating and cooling systems. Some codes let you use less insulation in one area provided that you compensate for the deficiency by increasing the insulation in another area.

Drawings

Generally, two sets of **design drawings,** also called **plans** or **blueprints,** must be submitted to the town engineer for his review and approval. One set is returned to you with the engineer's comments and required changes marked in red, and the duplicate set is kept in the files of the building department. If the changes are minor, the engineer stamps the drawings **"Approved as Noted,"** and a resubmittal is not necessary. On the other hand, if the changes are extensive, you may have to resubmit two revised sets for approval. The **originals** or **tracings** must be changed to reflect the engineer's comments before copies are made for construction. The approved drawings are an important document and must be kept in a safe place.

The blueprints must be clear, to scale, and include all the details required to contract and construct the house. A set of drawings must include:

1. A foundation plan showing the layout, dimensions and specifications of the foundations and the basement if there is one, the location of the stairs leading to the main floor, the location of the boiler and water heater and the sewage pipes.

2. Floor plans, one for each floor showing the floor layout with each room identified and fully dimensioned, the stairs leading to the other floors, the location and dimensions of the closets, the size and spacings of the floor joists, the type and size of the flooring, and the location of electrical outlets.

3. Longitudinal and lateral cross sections showing the foundations, floors, roof, and insulation.

4. Detailed drawings showing the kitchen cabinets, vanities, fireplaces, beam

ceilings, window and door schedules, plumbing layout diagram, etc.

5. Exterior elevations of the front, back, and sides of the house, showing the exterior siding, roofing, windows, exterior doors, and final grade elevations.

Land Survey and Staking Out the House

A land survey of a house is a scale drawing showing the boundaries of the lot, the surrounding area, and the position of the house and all other improvements such as driveways, walkways, deck, porch, and swimming pool. The survey must be made by a **state-licensed land surveyor** who is legally responsible for defining the property lines accurately and insuring that the house is being built on the correct lot. The surveyor must be familiar with the local zoning ordinances, particularly those establishing the minimum setback, right- and left-side yards and rear yard.

The first step is to clear the building site of trees and other obstacles. Next, retain (hire) a land surveyor and give her your name and the lot and block numbers. Subsequently, she visits the county office where the deed is recorded and obtains the legal description of the lot and the deed restrictions, if any. After doing this, she sets an appointment to meet you at the lot to define the property lines and **stake out the house.**

Usually, the surveyor works with a team of two or three assistants. Her tools consist of either a **theodolite** or **transit**, one or two **leveling rods, tapes,** 2 × 2-inch stakes, nails, and hammers. A theodolite is an instrument consisting of a telescope that can rotate around horizontal and vertical graded circles to measure horizontal and vertical angles with great precision. A transit is a similar instrument. A leveling rod is a frame made mostly of wood and graded in feet and decimals, or meters and decimals.

First, the surveyor stakes the property lines by inserting 2 × 2-inch stakes with colored ribbons in the ground at the corners of the lot and along the property lines. Next, she explains to you the front setback, right-side yard, left-side yard and rear yard, which determine the boundaries within which the house can be built. It is then your decision to tell the surveyor exactly where you want the house to be positioned. Generally, it is preferable to construct the house as close as possible to the street to shorten the lengths of the driveways, water, sewer, gas, and telephone lines, and to increase the depth of the back yard. It is also preferable that the house be built on a high spot to minimize the adverse effect of underground water. Besides, a house built on high ground looks bigger and better than one built on low ground.

After you tell the surveyor where you want the house to be located, she begins staking out the house. This is the process of establishing the **building lines,** which are the outside edges of the exterior walls of the house. First, she inserts initial stakes at the corners of the house (Figure 17.1). If you approve the location, she checks that the house will be perfectly square by measuring the diagonals AC and BD and ensures that they are equal in length. She then proceeds with setting **stakes** and **batter boards** at 15 to 20 feet from the building lines to clear the excava-

tion. (The initial stakes will eventually be removed when the ground is excavated.) The top elevation of all the batter boards should be the same so that it can be used as a benchmark in establishing the top elevation of the footings. Next, nails are hammered into the top of the batter boards such that any line stretched tightly between the nails of opposite boards precisely defines a building line. Once set, the stakes, batter boards and the nails must not be disturbed, moved, or covered until the foundations are constructed.

Afterward, in her office, the land surveyor prepares the preliminary survey drawing of the lot showing the building lines of the house and a brief description of the building: one-story frame dwelling or two-story frame and brick dwelling. She states the date on which the survey was made. This preliminary survey does not have to include the driveways, decks or any improvements other than the house. These items can be added in the final survey. She then seals and signs the survey and mails it directly to the town inspector or engineer. Naturally, she mails you a copy for your records. A copy of this survey is required for obtaining a construction loan.

Work Sheet for Heating and Cooling

A work sheet for heating and cooling includes the calculations of heat loss from the house in winter, and heat gain in summer if the house has central air-conditioning. By using the outside and inside heating and cooling design temperatures provided by the local building department, the heat

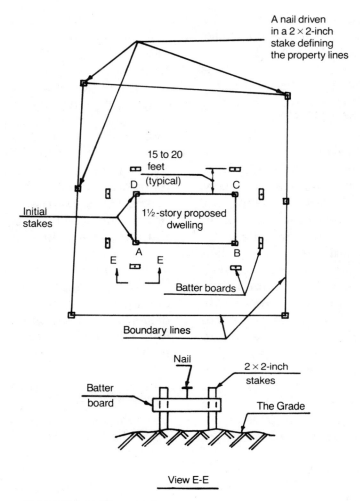

Figure 17.1 Staking out a house.

loss or gain through the exterior walls, windows, exterior doors, ceiling and floors due to conduction and infiltration is calculated based on the actual amount of insulation called for in the drawings. The **R-values** (see Chapter 31) indicating the resistance of the insulation must be equal to or exceed the values required by the state's energy conservation code. Usually, the work sheet is prepared by a **heating engineer.** However, a layperson can be trained to do it in a short period of time. A helpful guide in this regard is *Handbook of Fundamentals,* published by the **American Society of Heating, Refrigerating and Air-conditioning Engineers** (ASHRAE).

The capacity of the heating boiler or furnace is based on the amount of heat loss as calculated in the work sheet. Some codes specify that the rated capacity (output) of the heating unit must not exceed 125 percent of the calculated heat requirements. A bigger unit than you need will be energy inefficient. Similar calculations are made to size the central air-conditioning unit(s) based on the inside and outside design cooling temperatures, as given in the local code.

Heat Sheet

The heat sheet is a drawing similar to the floor plan, showing the layout and sizes of the ducts, the registers, and returns for air heating and cooling; or the layout and sizes of the feed and return pipes, baseboard heaters and radiators for water heating. It also shows the location and capacity of the boiler or furnace.

The heat sheet should be prepared by a heating engineer to ensure that the entire house is heated or cooled evenly. If the regulations of the local building department require that the plumber applies for an independent permit, it will be his responsibility to hire a heating engineer to prepare the heat sheet and obtain the approval of the town's engineer or inspector.

If the design calls for air ducts, the plumber should coordinate his work with the rough carpenter in establishing exactly where the ducts are to penetrate the floors in order not to injure the structural integrity of the framing. Wall ducts usually run between the studs.

Soil Percolation Tests

If the household sewage is to be disposed of on site, **soil percolation tests** must be performed to determine if the soil is able to absorb treated sewage. The soil must pass these tests before a building permit is issued. The number of tests depend on the planned system of sewage disposal, whether it is an **absorption field** or a **seepage pit.** Both systems are explained in Chapter 39. The test results are a factor in determining the size of the absorption system, depending on the type of soil (see Chapter 19).

The tests must be conducted under the supervision of an engineer or inspector from the department of health of the state. Each state has its own test procedure but the objective of the test is the same: to determine the rate of percolation of the soil when it is saturated.

Absorption Field   If an absorption field is planned there must be at least four feet of porous soils above rock, impermeable soil, or high seasonal ground water. In many states, the latter occurs in April and May due to annual snow melt combined with heavy spring rain. Some states require that at least two percolation tests be performed within the area of the absorption field. Other states require that the soil in the absorption field must be consistent. This is determined by digging two holes about ten feet deep and fifty feet apart in the area of the absorption field. If the soil is found to be homogeneous, one percolation test is performed in between the two holes.

Seepage Pit   If seepage pits are planned, two percolation tests should be performed at the proposed site of each pit, one at half the pit depth and the other at the full depth. If different soil layers are encountered in the test pit, a percolation test may be required for each layer and an average percolation rate based on the depth of each layer is calculated.

Site Investigation   Before the test begins, the engineer or inspector conducts a site investigation or **field survey** whereby he walks around the proposed site of the sewage disposal system. He checks that the minimum separation distances between the system and water wells, buildings, and property lines are maintained. (Where drinking water is obtained from shallow wells and where the soil is coarse gravel, seepage pits are not allowed.) He also checks the grade. Slopes greater than 15 percent are too steep and should be avoided. The engineer also looks for clues that may indicate the soil formations. For example, rock outcroppings warn that shallow soil exists and may serve to indicate the direction of ground water flow.

The Percolation Test   The percolation test must be performed according to the procedure established by the state. Generally, the test begins with digging a square hole approximately 12 inches square. If an absorption field is planned, the depth of the hole should be 24 to 30 inches below the final grade elevation. If seepage pits are planned, two holes are dug to depths equal to one-half and the full estimated depth of each seepage pit, respectively. The depths of the seepage pits are determined based on the characteristics of the soil. In order to prevent the soil from heeling, or caving-in, a larger excavation may be made at the top portion of the hole provided that the lower portion is kept vertical. A measuring stick is inserted in the hole (Figure 17.2) to measure water levels during the test. The stick is secured in its place by a batter board. Next, two inches of gravel or small stones are placed in the bottom of the hole to reduce scouring and silting action when water is poured in the hole. It is important not to disturb or move the batter board during the test.

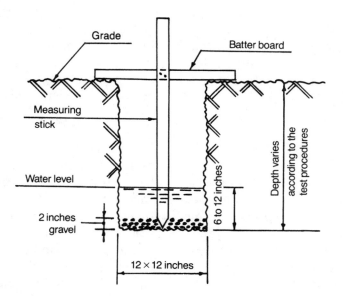

Figure 17.2 Soil percolation test.

The following step is to saturate the soil by pouring clear water in the hole to the depth specified in the test procedure. Water is periodically added to compensate for the water that seeps away. The period of saturating the hole varies between 30 minutes and a few hours, depending upon the state. After the saturation period, any loose soil that has fallen in the hole is removed.

Finally, clean water is poured into the hole with as little splashing as possible to the depth specified in the test procedure (between 6 and 12 inches). The time it takes the water to drop is measured in minutes-per-inch for a few inches. This measurement determines the rate of percolation of the soil and the size necessary for the absorption field or seepage pits. It should be emphasized that the details of the test procedures vary considerably from state to state, and you should contact the department of health of the state to get the exact test procedure for your area.

Well Water Test

If municipal water is not available, a well has to be constructed to obtain the water needed for domestic use. Well water has to be tested by the department of health to insure that it is clear, safe to drink, and does not contain significant amounts of objectionable chemicals. Each state establishes its own test procedures.

The production of the well should be sufficient to meet the household demand. It is preferred that the well yields at least five gallons-per-minute. The water must pass the **physical**, **bacteriological**, and **chemical tests** before a building permit can be issued.

General Sanitary Survey First, the health department inspector tours the area in what is known as a general sanitary survey. The purpose of this survey is to check that there is no nearby source of contamination such as an absorption field, underground chemical storage, or gasoline leakage.

Water Testing A sample of the water has to be tested in a laboratory approved by the department of health. The purpose of the physical test is to insure that the color, taste, and odor of the water are within certain limits. The bacteriological test checks the existence of the coliform group of bacteria that is a sign of contamination. The purpose of the chemical test is to insure that the amount of certain chemical substances does not exceed what is allowable by the health department. High iron content causes brown staining on laundry and plumbing fixtures. It also causes bitter tasting tea and coffee. Manganese causes black spots on laundry. Biocarbonates and sulfates cause water hardness, which hampers the water's ability to produce suds, and causes deposits on the walls of the water pipes and water heater.

If the content of any substance exceeds the allowable limit, it does not mean that the well is automatically rejected, but that the water needs treatment. The testing laboratory usually recommends the required type of treatment to make the water suitable for domestic use.

Building Permit Application Form

Each locality establishes its own building permit application. A typical form includes the following information:

1. A statement that you (the applicant) will comply with all provisions of the building code in the erection of the building whether specified in the application or not.

2. The location of the lot identified by the section, block and parcel numbers and by address.

3. Zone classification: residential—the number of the units per dwelling should be specified (e.g., one-family house).

4. Size of the lot: front (in feet); rear (in feet); and average depth (in feet).

5. Area of the lot (in square feet).

6. Percentage ratio between the area of the building and lot area (some zonings require that the house should not be built on more than a certain percentage of the area of the lot).

7. Cubic content (size) of the house: The area of each floor is multiplied by its height, then they are all added up. (Some building departments base the application fees on the cubic size of the house.)

8. Estimated cost of the building (in dollars).

9. Brief statement covering the work planned, e.g., "one and a half story building of frame construction with half-basement." (The town engineer can help you with this item.)

10. Number of stories (many localities restrict the number of stories).

11. Height from curb level or grade level to the highest point of the roof (in feet).

12. Front setback (in feet).

13. Side yards: right (in feet), left (in feet).

14. Rear yard (in feet).

15. Character of ground (rock, clay, sand, fill, gravel, ashes).

16. Name and address of the owner.

17. Name and address of the architect, if any.

18. Name and address of the builder, if any.

19. Name and address of the plumber (many localities require that the plumber apply for a separate permit and be responsible to the town engineer directly. However, you may apply for a permit before selecting a plumber).

20. Name and address of the electrician.

21. Workmen's Compensation Insurance Policy, with the name of the insurer, the policy number and the expiration date.

22. Signature of the applicant.

Before the building permit is issued, you are required to pay a building permit fee. At least part of this fee is used to cover the cost of inspections.

Inspection Schedule List

Each locality establishes its inspection schedule list. The number of items subject to inspection vary considerably from locality to locality. Urban areas have a stricter inspection schedule than rural areas. Here is a typical urban inspection list:

1. Staking out the house.

2. The sewer lines and the connection to the public sewers.

3. Before pouring the footing.

4. After the foundations are built and before backfilling.

5. Checking the rough framing as it goes up.

6. The fireplace.

7. The rough plumbing.

8. Exterior siding and roofing including flashing.

9. The rough electrical.

10. The rough heating.

11. The insulation.

12. The fire protection around the boiler.

13. The chimney.

14. The handrails.

15. A spring to keep the door between the garage and the house closed.

16. The drywall or sheetrock.

17. Finish plumbing and plumbing fixtures, and finish heat.

18. Finish electrical wiring.

19. Flooring.

20. Septic tank and absorption field, if any.

21. The house trap connected to the sewage pipe before it leaves the house (see Figure 29.1).

22. Water well.

23. The final grade.

CHAPTER 18

Choosing and Dealing With Contractors

*I*n building your house, you will have a lot of dealings with contractors. Some time before construction starts, you will have to decide whether you will hire a general contractor to build the entire house or subcontractors to build parts of the house. Your decision to subcontract your house should be based on your experience, willingness and determination to get fully involved, knowledge of construction, and the time you can personally allocate to what is a major undertaking. If you do not feel up to it, it is better to give the job to a general contractor. But if you do decide to go ahead and hire subcontractors, then you should ask yourself: How much am I going to save? and is it worth it?

The answer to the first question is to give the drawings and specifications to different general contractors and get their prices. In the meantime, get estimates from subcontractors for various parts of the house. Add all these together plus a few percentage points to cover miscellaneous expenses such as architectural or engineering consultations. Compare both figures, and you'll get a very good idea of how much money you can save. (Generally speaking, the saving as a percentage of the total cost of the house is greater for big houses than for small ones.) Obviously, this is an elaborate process that may take a month or more to complete; you'll want to start working on it as early as possible.

The answer to the second question lies in the amount of savings. If it is small, then it may not be worth it unless you really want to be involved in the details and finishing. But if the savings are substantial, then you will have to give the matter serious consideration. Let us assume that the saving is estimated to be $30,000, then ask yourself: how much overtime do I have to work to net $30,000 after taxes? The next question you ask yourself is: do I have the $30,000? Analyzing the situation should help you make the right decision.

The availability of funds is another

consideration. A general contractor is going to build the house quickly and expects to be paid as soon as the work is completed. You must have practically all the funds available when construction starts. On the other hand, if you subcontract the house, you can build it at a slow pace. This makes a difference if your income is more than your expenses and you can allocate the difference to building costs. Further, you can build as much as is needed to get the certificate of occupancy and complete the house after you move into it. This gives you the leeway to build yourself a bigger house than your current financial resources allow.

Knowledge and experience in construction can be partially accumulated by watching the construction of other houses. If you are really interested, you can stop at any construction site and chat with the workers. Generally, you will find them willing to talk to you.

Reading about it is an important way to gather information that will aid you in managing the construction of your house. The more you read, the more you are prepared. They say in the military, "the harder the training, the easier the battle." You should start studying long before construction starts and read only the books that are practical and to the point.

Great care should be taken in choosing the general contractor since the success of the entire project depends upon the competence, managerial capabilities, financial soundness, and manners of one person. He buys the materials, hires the subcontractors and sees that they follow the drawings and specifications. He should be able to give directions and solve any problems arising during construction. The assumption is that you do not know

much about construction. On the other hand, subcontractors expect you to be knowledgeable about construction. If they have a problem, they want you to give an opinion. You should also be able to check the quality of their work and that they follow the drawings and specifications.

Choosing and Dealing With General Contractors

The first step in choosing a general contractor is to get a few names. Look to these possible sources:

1. Houses under construction.
2. Material suppliers.
3. Friends and relatives who have recently built houses in the area.
4. Your accountant, lawyer, or real estate agent.

After you get the names and telephone numbers of several contractors, call them to see if they are available. You should keep looking until you get three or four who are interested. At this early stage, and before you spend too much time or energy, contact the **Better Business Bureau, Chamber of Commerce**, and the **Department of Consumer Affairs** to learn whether there are complaints filed against these contractors. This gives you a clue, although nothing conclusive, about the reputation of the contractors. This is because these organizations report the number of complaints filed against a contractor but they do not verify the validity of the complaints.

The following step is to interview each contractor separately. You should show each one the drawings and specifications and ask for his opinion, comments, suggestions for modifications, and his reasons for changes. He may have some good ideas, but do not accept his suggestions or recommendations automatically on the assumption that he is the expert. If you do not have previous knowledge and experience in construction, have an architect or engineer with you during the interviews. All contractors should bid on exactly the same drawings and specifications, in order for you to make a meaningful comparison between prices. If you make any change after one or more contractors were given the drawings, you should call them and inform them of the change.

Ask each bidder to show you some houses that he is currently building or that he has already built (sometimes the owners allow you in), and ask him to give you the names and telephone numbers of their owners. Do not hesitate to call and ask them if they are satisfied with the contractor's quality of work and character. Also ask how prompt he was in fixing problems that arose after they moved into the house, such as windows that did not open or close freely or drywall that cracked.

Also ask each bidder for bank references. If he is financially sound, he should have no objection to providing you with this information. But if he refuses to furnish such information, it should be a cause for concern.

Some bidders prefer to be contracted on a "cost plus basis" whereby they charge you for the cost of material and wages of every person who works directly on the job, and add a percentage to cover their profit and overhead expenses (rent, secretary, telephone, electricity). This method is not to your advantage because you will have little control over what is being charged to you. Additionally, you cannot determine how much the construction will finally cost. The final cost may be beyond your budget. The contractor is likely to go only as far as your money goes, leaving you with an incomplete house. The recommended method is on lump sum basis whereby the bidder gives you a fixed price based on a specific set of drawings and specifications. There should be provisions regarding unforeseen problems such as may turn up during excavation.

It takes two to four weeks to get an estimate from a serious bidder. If he takes longer than that, it is a sign that he is either too busy or not serious enough. If a contractor's price is too low it may be that he either has made an arithmetic mistake in his bid or that his work is of poor quality. There are instances where a contractor bids low (but not *too* low) just to keep himself and his crew busy. There is nothing wrong with that. If you have any doubt about the capability of the contractor to complete the project as agreed upon, you can ask him to post a performance surety bond. You should be prepared to pay for the bond's premium, however.

You are not required to open the bids in a formal meeting. This makes all the bidders nervous. They may even resort to the technique of giving a low price with a long list of terms and conditions to compensate for the low price. If you feel that you cannot decide on which bidder gets the job, seek the help of an architect or a structural engineer. Further, you should state in advance that you have the right to reject any or all the bids without giving reasons. If you feel that the bids are too high, you may cancel them all and put the job up for rebidding.

One important factor that must not be overlooked is the personality factor. You must get along well with the contractor. Notice whether he is a good listener, gives you a reasonable amount of his time, and answers your questions. If you feel that his attitude is "Hey, I'm a very busy guy, I have several bigger jobs; I can't afford to spend too much time on your job," you are better off hiring another contractor.

After you choose a general contractor, you will both have to sign a contract as explained in Chapter 3.

As soon as the contractor breaks ground (starts excavation), it is natural that you and your family should become frequent visitors. Mostly you will observe, asking questions and watching the house grow. If you find a mistake or anything you do not like, talk to the contractor or his representative, but never to the worker responsible. Occasionally, you will be called upon to choose the colors, but not the types of the tiles or the paint. Any deviation from the contract triggers an "extra" and the contractor will not hesitate to flag the dollar sign. If any problem arises, try to solve it quietly. Yelling and screaming are counterproductive and must be avoided.

Choosing and Dealing with Subcontractors

Subcontractors are specialized contractors, such as plumbers, masons, carpenters, roofers, etc. There are about 40 trades involved in the construction of a house, and you may need to hire a comparable number of subcontractors. Hiring subcontractors is less formal than hiring a general contractor, but takes more time.

Local subcontractors are familiar with the area and the local building and energy conservation codes. They are likely to give you competitive prices, since they do not have to commute a long distance. The best way to find subcontractors may be:

1. Through material suppliers: Lumber suppliers can provide you with names of carpenters; brick suppliers can give you the names of masons, and so on.

2. Through other subcontractors: A carpenter may know a good plumber who worked with him on a previous job.

3. Through neighbors who have recently built their houses.

4. From houses under construction.

Contractors will approach you or leave their business cards for your consideration when they see your house going up.

Most of the subcontractors you will be dealing with are not well known. They get most of their work through word of mouth. They run the business from their homes with their spouses taking telephone messages and often handling the bookkeeping. This reduces their overhead expenses and makes them competitive. Each subcontractor maintains a small crew that varies in number according to work load. When one gets a big job, he or she borrows additional workers from another not-so-busy fellow contractor and vice versa.

In dealing with subcontractors, it is important to respect them for their abilities and what they know. Each one of them has spent most of his or her adult life learning the trade whether it is carpentering, masonry, plumbing, or electrical work. They develop a feeling for how to do the job efficiently, how long it takes and how much it costs. Many subcontractors can tour the building site with or without you, take some measurements, make a few calculations and determine the price and time required to complete the job in a very short time.

Naturally, not every subcontractor is topnotch. Some are average, others are below average. You are therefore obligated to ask the bidders for references' names, addresses, and telephone numbers. Do not hesitate to check the references and ask about the quality of the bidder's work, and his or her character and reliability. You should also find out whether you would be dealing with a troublemaker, or someone in the habit of taking his or her clients to court, or attaching liens to their houses.

Different subcontractors will give you different prices for different reasons, some legitimate, others not. If a certain subcontractor has a reputation for good workmanship, he is likely to ask a higher price than a less reputable subcontractor. Another factor that has no relation to workmanship is whether a subcontractor is busy. He will raise the price because he does not really need, or will find it hard to handle your job. If he gets your job at the high price, he is likely to put his other jobs on hold and start immediately on your project.

Unfortunately some subcontractors will take you for a ride if they can: While I was building my house, a carpenter gave me a price for rough and finish carpentry that was three times the price of the successful bidder. His reason was that he would offer to act on my behalf as a superintendent should problems arise from the other contractors. I looked at him in disbelief and shook my head. Here is a man who wants to provide me with unneeded services at a huge cost, no doubt planning himself to cause the very problems he was offering to fix. I had no choice but to politely tell him to take a walk. To frustrate such an attempt, you must solicit bids from at least three subcontractors. The prices of the serious ones will be close. The ones that are too high or too low must be eliminated.

The attitude of the subcontractors will give you a clue as to their willingness to work for you. If a subcontractor does not return your telephone calls, it means that he is not eager to bid on your job. He may be busy, pretending to be busy, or perhaps he does not want to work for you because something you did or said offended him. Before the stairs of my house were installed, the carpenter laid two boards spanning the staircase for the workers to walk on. One bidder chose to stand on the boards while discussing his bid with me. I asked him to step off the boards for his own safety, but he would not listen to me. In the heat of discussion, he wanted to dramatize his point of view and started jumping up and down on the boards. I was horrified, particularly since the man weighed about 280 pounds, and I asked him again to get off the boards, otherwise he might break them and fall into the basement. Apparently, he was offended and left. I never saw or heard from him since. It did not matter to me: People's safety overrides any other consideration. Had that fellow gotten injured, he would have sued me for negligence. The point is that if a subcontractor is not interested in bidding on your job, look for another one. Also note whether the bidder comes to your appointment on time. A person who arrives late without a *very* good excuse is unreliable.

Conversely, if a contractor bugs you a lot by telephone, it may indicate that he is not busy and there may be something wrong about him. Once I had such an experience, and I decided to interview the guy. In the conversation, he complained that his clients are no good because they do not pay him for his work. He was portraying himself as the underdog, an easy prey whose clients take advantage of him. He was driving a big new car, so naturally I asked him how he could afford such a car if his clients do not pay him. His answer was that it was "a birthday gift from his father." I became very apprehensive that his story was bait, and I did not bite. I decided not to deal with him. Several months later, I met an ex-partner of his and sure enough, I learned that he was in the habit of doing lousy work, then attaching mechanic's liens on his client's houses when they did not pay his asking prices.

Subcontractors like to chat with you. They are curious and want to know your background and how much you know about the work in progress. They are interviewing you while you are interviewing them! It is, therefore, imperative that you learn as much as you can about the subject at hand before you talk to them. If you impress them that you know the secrets of their trade and how to check on their work, they will give you a reasonable price be-

cause they know that you will solicit estimates from their competitors. And whoever gets the job will perform good work for you.

If a subcontractor makes a mistake your attitude toward him should be: Let us try to solve the problem together. Be calm and do not lose your nerve. Contractors big and small make mistakes, but they correct them. You first ask him: What do you suggest to solve the problem? If both of you cannot find a satisfactory solution, ask him again: Who can solve the problem? Depending on the problem, you may have to consult a professional engineer or an architect.

It is very important to keep a sense of humor. Most subcontractors and construction workers appreciate that very much. A little joke at the right time can open up clogged communication channels and ease a lot of tension. One of the subcontractors who worked on my house was a karate expert. Before he completed his work, he sent me a bill for the full amount. I gave him a call to clarify the situation. His wife answered the telephone and told me that he was taking a nap. I asked her to please tell him to call me when he wakes up, and she promised to do that. The man never called. Three weeks later, he mailed me the same bill for the full price once more. I checked the house and found that his work was still incomplete. I was irritated and decided to give him another call. His wife answered the telephone. When I asked about her husband, she told me that he was taking a nap. I became frustrated and decided to vent my anger in a joke. I said to her in a quiet tone: "I called three weeks ago and you told me that he was taking a nap. Again you are telling me that he is taking a nap.

Well, I hope it is not the same nap!"
The lady exploded in hysterical laughter. She repeated in a loud voice "no it is not the same nap—no it is not the same nap." Apparently, her husband was awake and she wanted him to hear the response to his "taking a nap" ploy. The karate expert was hiding behind his wife and I gathered from her loud laughter that she was not happy about that. In any case, the joke brought results, and the man completed the work the following day.

CHAPTER 19

Types of Soils and Excavation

Soils are a mixture of boulders, gravel, sand, silt, clay, and organic matter. They support the house, shape and form the final grade and dispose of the household sewage if the house is not connected to municipal sewers. To perform these functions satisfactorily, soils have to have good bearing and permeability qualities.

Types of Soils

As mentioned earlier, soil maps that show the types of soils, their general characteristics and suitability for various uses have been developed for many parts of the country. These maps are available at the **U.S. Soil Conservation Service.** Useful information may also be found in the maps and reports of the **U.S. Geological Survey** and the **U.S. Department of Agriculture.** These maps and reports, if available, are helpful in determining the general soil characteristics but they are not substitutes for on-site soil investigation.

The simplest method of on-site inspection is to dig a hole or several holes within the building site and the absorption field of the household sewage disposal system. A backhoe can dig these holes quickly. This will reveal the soil layers, the depth and composition of each layer, how easy the excavation will be, and the depth of the subsurface water. The most common types of soil are:

Boulders Boulders are rocks that are over 8 inches in size. They are usually embedded in other types of soil. If big boudlers are in the way of pouring the foundation or laying the sewer pipes, they should be removed to avoid creating hard spots which would cause the foundations or sewer pipes to crack. Their place should be filled with clean sand or gravel. If the boulders are too big to be removed, the foundations will have to be reinforced with steel bars. The design of the reinforcement must be determined by a structural engineer.

Gravel   Gravel is rock fragments or particles varying in size from $\frac{1}{8}$ inch to 8 inches. Gravel is good for both bearing and absorption. However, it is rare to find soil consisting of gravel only.

Sand   Sand consists of rock particles or grains of sizes varying from 0.05 millimeter (1 inch = 25.4 millimeters) to $\frac{1}{8}$ inch. Sand is a good material for bearing and absorption. Care must be taken that the sand layer is deep and not just a thin layer underlain by weak soil. Very fine sand should be evaluated with great caution since water may convert it into quicksand causing the foundations to fail. A geotechnical or structural engineer must be consulted.

Silt   Silt consists of rock particles that are medium in size between sand and clay, that is from 0.005 to 0.05 millimeters. Silt exists in two forms: inorganic and organic. **Inorganic silt** has little plasticity, meaning if you add water to it, the paste does not deform without cracking. **Organic silt** is silt mixed with particles of organic matter such as decayed vegetables or shells. Its color ranges from light to dark gray depending on the contents of organic matter. Silt is not very good in bearing although it can support the weight of a house. Its absorption characteristics are poor.

Clay   Clay consists of very fine particles less than 0.005 millimeter in size. It is plastic when wet and hardens and cracks when dried. A ball of moist clay can be formed into a long ribbon without breaking. Clay soil may be stiff, medium, or soft depending on its natural consolidation. When loaded, stiff clay takes years to settle. This is because it loses its moisture content very slowly. Therefore, if you build your house on stiff clay, you should expect the plaster or drywall to be cracking for a few years, but this does not mean that the house will fail. Clay is good in bearing but very poor in absorption.

Loam   Loam consists of a mixture of sand, silt, and clay. When dry, it forms a cast that can withstand careful handling. When wet, it is partially plastic, and forms a cast that can be handled freely without breaking. Loam has some bearing and absorption capabilities.

Sandy Loam   Sandy Loam is mostly sand, but has some silt and clay in it. Individual sand grains can be seen and felt. When dry, if squeezed in the hand and released, it falls apart. But when wet, it can stand handling without breaking. Sandy loam has some bearing and absorption capabilities.

Silty Loam   Silty loam includes a large percentage of silt and a small percentage of sand and clay. Natural lumps can turn into soft powder with little pressure. It has considerable plasticity when wet. When moist, lumps can be handled without breaking. Its bearing and absorption capabilities are poor.

Clayey Loam   Clayey loam includes a large amount of clay and a small amount of sand and silt. When moist, it has a smooth texture and can withstand considerable handling; when dried, it becomes hard. Clayey loam has bearing capability, but is poor in absorption.

Peat Peat consists of fiberous, partially decomposed vegetable tissue. It ranges in color between brown and black. Peat is compressible and thus must not be used for bearing. Its absorption capability is also poor.

Adobe Adobe is a type of plastic clay found in the southwestern United States. It is poor in both bearing and absorption. It is used in building because of its good thermal insulating characteristics.

Till Till is a glacial deposit of clay, silt, sand, gravel and boulders. It is found in the regions that were glaciated during the ice age. Undisturbed compact till is good in bearing, but loose till should be carefully investigated. Till has absorption capability.

Excavation

Excavation is the removal of the soil in order to construct the house's foundations and basement, lay the sewage and water pipes, gas and telephone lines, and install the sewage disposal system and swimming pool, if any.

Clearing the Site Before excavation starts, and even before staking out the house, the construction site must be cleared of trees. The trees should be cut three to five feet above the ground depending on their size. This is to allow the excavating machine to pull the stumps out of the ground.

Depth of Excavation The depth of excavation is determined by: (1) the design drawings, (2) the depth of good

soil, and (3) the elevation of the public sewer line.

1. The design drawings should show the depth of the footings and basement with respect to the final grade around the house. If the house has a basement, the footings are poured below the floor slab of the basement. If the house does not have one, the footings should be poured below the frost line, but in no case less than three feet deep in order to reach undisturbed soil. The **frost line** is the depth below the surface of the earth in which the ground freezes in winter. Soil retains moisture either from rain, watering the lawn, or underground water. During winter, the moisture in the soil *above* the frost line freezes into ice causing the soil to swell and heave. During spring, the ice melts causing the soil to shrink and settle. Soil swelling and shrinking is not uniform and if a house is constructed above the frost line, its foundation and walls will crack and the cracking will increase with every freeze–thaw cycle.

2. The excavation must go deep until good soil is reached. Good soil means soil that can sustain the weight of the house throughout its life without exhibiting any appreciable differential settlement that can cause foundation and wall cracking. What constitutes good soil is determined mainly by experience. The soil classifications explained earlier in this chapter provide you with theoretical background, but this knowledge is not a substitute for an expert opinion. In most cases, an experienced excavation contractor can determine what is good soil. In all cases, you must consult with a geotechnical, structural, or town engineer. And as mentioned before, it is a good idea to ask the neighbors about the depth and type of foundation of their houses, and whether they have any cracking problems, for this may give you a clue regarding the soil con-

ditions in the area. However, you must keep in mind that soil conditions may vary markedly from lot to lot.

3. The elevation at which the house's sewer pipe is connected to the public sewer line is another factor that determines the depth of excavation. Sewage moves by gravity. Thus the elevation of the house's sewer pipe at its exit from the house must be higher than the point at which it connects to the public sewer line. The grade or rate of slope of the sewer pipe should be about ⅛ inch per foot. The sewer pipe should enter the house below the footing and the concrete slab of the basement. The elevation of the bottom of the footing should be about eight inches higher than the top of the pipe to give the plumber room to set the sewer pipe straight and to secure it in place by pouring concrete around it (Figure 19.1).

Machine excavation should stop at about three inches above the final elevation of excavation. These three inches should be excavated manually to eliminate soil disturbances caused by the excavating machines. Leave room around the excavation to give the concrete trucks access to the pouring area.

Width of Excavation   The width of the excavation should extend at least two feet beyond the outer edges of the footings to give construction workers room to build the foundations. The edges of excavation should be sloped outward as shown in Figure 19.2, to prevent soil from sliding into the construction area.

Trenches for Footings   It is normal practice in excavating for a house with a basement to excavate the area of the basement first, then excavate trenches for the footings (Figure 19.2). The width

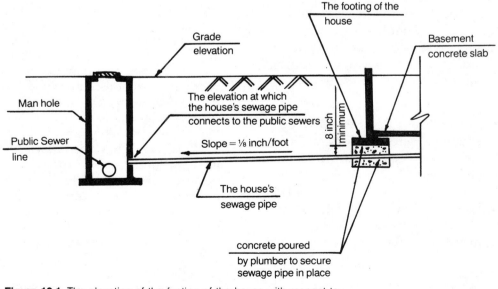

Figure 19.1 The elevation of the footing of the house with respect to the elevation of the sewage pipe.

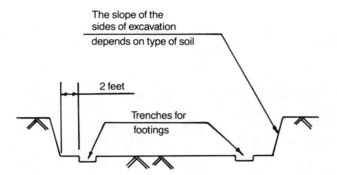

Figure 19.2 Profile of excavation for basement and footings.

and depth of the trenches are equal to those of the footings. Footings can then be poured in the trenches. This technique saves the side forms of the footings and reduces the volume of excavation. However, this is not recommended if subsurface water exists because the drainpipe (Chapter 21) must be laid at the bottom of the footings.

Water Drainage If subsurface water is encountered where the footings are to be poured, water will have to be drained. This must be done carefully in order not to disturb the soil. First, a sump is dug where it will not interfere with construction. Second, small trenches connecting the areas to be drained to the sump are dug. Third, the drained water is pumped out of the sump. It is extremely important that the drained water not include any soil particles; this disturbs the soil and is thus detrimental to the safety and integrity of the foundations. If the pumped water starts to draw soil particles with it, pumping must be stopped immediately, and a geotechnical or structural engineer must be consulted.

Excavating Machines

The three machines used most in house excavation are: (1) the bulldozer, (2) the backhoe, and (3) the front end loader.

The Bulldozer The bulldozer is a tractor with a front mounted blade that can be raised or lowered. It is used for site excavating, grading and backfilling. Most bulldozers are of the **crawler** type, meaning they move **on tracks** as the machine shown in Figure 19.3. They are powerful and can move big boulders and tree stumps, and dig hard soils. The tracks help them to get a better grip on slippery or muddy soils and to move up and down steep slopes. Some bulldozers are mounted on four-wheel-drive rubber tires.

The Backhoe The backhoe has a bucket at one end and a hoe at the other (Figure 19.4). Both the bucket and the hoe can move hydraulically in different directions. The end of the machine facing the hoe has two opposite struts that can extend diagonally outwards (shown in Figure 19.4 behind the big wheels) until they reach the ground and lift the machine off its

Figure 19.3 The bulldozer.

Figure 19.4 The backhoe.

wheels. This gives the machine great lateral stability. Most backhoes are mounted on four wheels, but some are on tracks. The backhoe is a versatile machine. It is used for digging trenches, grading, loading, moving earth, back-filling, and spreading top soil.

The Front End Loader   As its name implies, the front end loader has a mounted bucket at its front that can be hydraulically raised, lowered and tilted. The front end loader may be mounted on four wheels as shown in Figure 19.5, or on tracks. Its bucket is larger than that of the backhoe. The bucket's capacity ranges from one to four cubic yards for the machines used for house excavation. Front end loaders are used for digging large areas such as the basement or swimming pool. It can also be used to haul the soil from one place to another, to spread the soil, to load trucks and to backfill.

Contracting for Excavation

Excavation may be contracted on a lump sum basis or by renting the machines and their operators by the day. The rate depends on the size of the machine and the market condition. It is preferable to assign the excavation to an experienced contractor who can counsel you on the soil condition. Crawlers are not allowed to move on paved public roads, and therefore must

Figure 19.5 The front end loader.

be carried on transport vehicles at extra cost.

Before excavation starts, the contractor and machine operators should be made aware of the location of the stakes and batter boards marking the position of the house so that they do not remove, disturb, or cover them in the course of excavation. At the outset, you should instruct the contractor to strip and stockpile all the soil that can be used as top soil. Later, after grading is completed, this soil can be spread around before lawn seeding or sod-ding, or be used for a vegetable garden or flower beds. This saves you the cost of buying expensive top soil.

If the contractor encounters boulders or if the soil conditions are such that he has to go deeper or do more work than what was contracted, he should be fairly compensated. In the meantime, you should be aware that some contractors tend to exaggerate their claims for extras. When estimating the cost of extra work, try to proportion the amount of extra work to the amount of work in the original contract.

CHAPTER 20

Blasting

*I*f your lot is in a rocky area and solid rocks are in the way of the foundations, basement, sewer line trench, or swimming pool you may have to blast.

Blasting breaks the rocks into small pieces that can be excavated. The area to be blasted should extend a couple of feet beyond the outer edges of the foundations to provide space for the workers and equipment required for constructing the foundations.

At the outset the blaster should be made aware of the location of the stakes and batter boards indicating the position of the house, and he must not disturb, remove or cover these stakes while performing his work.

Seismographic Analysis

Before blasting begins, **seismographic analysis** is sometimes conducted on site to determine its vibration characteristics. A few blasts are detonated at various distances from a seismograph to determine the parameters of the **scaled-distance formula** used to calculate **particle velocity** at a given point. (The distance between this point and

the center of the blast must be known.) The particle velocity is indicative of the vibration intensity and is measured in inches-per-second (ips). The factors that affect the particle velocity at any point in a particular site are the rocks' characteristics, the weight and power of the explosives per detonation; and the distance between the point and the center of the blast.

In no way should the particle velocity at the nearest point of the nearest building, as calculated by the scaled-distance formula, exceed 2.0 ips in order to avoid excessive vibrations and high levels of noise. Blasters usually shoot for a particle velocity of about 0.60 ips in residential areas.

The blaster calculates the type and weight of explosives required to blast a cubic yard of rocks. Next, he determines the size (diameter) and spacing of the **blasting holes** in which the explosives will be placed, and the amount of explosives needed in each hole. The depth of the holes is equal to the depth at which the foundations will be constructed. The blaster then makes a layout sketch showing the location, diameter, and depth of the holes. Once

Figure 20.1 Drilling machines.

Figure 20.2 Covering the blasting holes with protective cone-shaped tar paper.

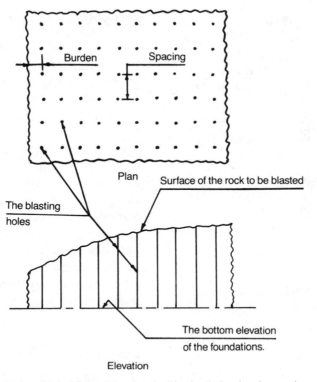

Figure 20.3 A layout showing the blasting holes, burden, and spacing.

Drilling

all this is done, drilling of the blasting holes begins.

Blasting holes are drilled with machines similar to those shown in Figure 20.1. These machines are propelled by compressed air that is also used to clean debris from the holes. As soon as the holes are drilled and cleaned, the blaster covers them with cone-shaped tar paper (Figure 20.2) to protect them from mud, dirt, and rain.

The distance between the edge of the rock and the nearest line of holes is called **the burden,** and the distance between two successive lines of holes is called **the spacing** (Figure 20.3). The burden should be less than the spacing so that the blast will throw the rocks in the direction of the burden. Blasting holes may be drilled vertically or at an angle depending on the shape of the final excavation and the nature of the rocks.

Loading and Stemming

Loading means wiring the explosive charges and inserting them inside the blasting holes. The explosives should be placed deep in the hole leaving the top portion of the hole free. This top portion is then filled with pea gravel or drill cuttings (Figure 20.4) and compacted, a technique called **stemming.** The purpose of stemming is to confine

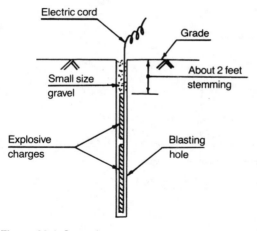

Figure 20.4 Stemming.

the forces of explosion and to prevent the accidental ignition of the charges before blasting or firing. The depth of stemming depends on the diameter of the holes and the distance of the burden. In residential blasting this depth is about two feet.

Delay Caps

A delay cap is a device that delays the explosion of the charges for a few milliseconds (1 millisecond = 1/1000 second) after the electric circuit is closed. The amount of delay is marked on each cap (e.g., 25, 50, 75, 100, 125 milliseconds). Using caps with different delays allows blasting of several holes in rapid succession in one firing. This technique increases blasting efficiency, reduces the burden, produces better fragmentation, and controls the throw.

Blasting

Blasting begins with the shallow end of the rocks, which has the least burden, then proceeds toward the other end. Several blasting holes are loaded and wired at one time. The charges in the holes are connected to caps with successive delays. After stemming, all the loaded holes are connected in an electric circuit. The source of the current may be a battery or a blasting machine. When the circuit is closed, the explosives are detonated one after the other according to the amount of each delay. The number of holes to be detonated at one time and the sequence of detonation and delays are determined by the blaster.

If the area to be blasted is one solid flat rock, two inclined holes are drilled in the middle of the rock and blasted a few milliseconds apart. Blasting then proceeds outward until completed.

Disposing of the Rocks

Disposing of the rocks resulting from blasting can be a problem. Trucking away rocks is very expensive. Therefore, you should bury as much as you can on site. Here are some suggestions:

1. Raise the final grade elevation around the house.

2. Use some of the rocks to construct a dry hole (or dry well) to drain the water of the drainpipe. The elevation of the bottom of the dry hole must be lower than the elevation of the bottom of the footing in order to keep the basement dry. The drained water will evaporate during hot weather.

3. Some rocks may be buried in soily areas in the lot. As a last resort, excess rocks are trucked away.

Safety Precautions

Safety precautions must be taken to ensure that blasting will not harm hu-

Figure 20.5 Steel mat being moved by a machine.

mans, animals, or properties. One of these precautions is to cover the area over the holes to be blasted with heavy steel mats, or curtains, to prevent rocks and debris from flying in the air. (A rock propelled by the force of a blast can kill a human being). These mats are so heavy that they require machines to move them around (Figure 20.5). For sufficient protection in residential areas, several mats may be needed. Figure 20.6 shows three mats laid on top of one another to cover the area of a six hole detonation. I observed this detonation and noticed that the mats contained the broken rocks and debris completely. The mats just bulged a little. A few years ago while

Figure 20.6 Three mats covering a six hole detonation.

I was interviewing blasters, one of them told me that he can blast very close to a window without breaking the glass!

Another safety precaution is to check the electric circuit before closing it (firing) to ensure that there are no loose connections or groundings. Such defects may cause misfirings that can be fatal.

A third safety precaution is to warn everybody to clear the site before firing. This must be done by well understood signals, such as three loud whistles. After firing, the blaster gives another type of signal to declare that blasting is over. The mats are then removed and the broken rocks and debris excavated and piled on the side.

Good Engineering Practice

Good **engineering practice** are beyond the requirements of the local building codes and national specifications. It is based on personal experience and, therefore, may differ from one engineer to another.

Good engineering practice suggests that you do not build a part of the foundation on rocks and the other part on soil. Soil settles with time (stiff clay takes years to settle) and heaves and settles with the rise and fall of subsurface water; rocks do not. If a part of the foundation is laid on soil and the other on rocks, differential settlement occurs between the two parts resulting in cracking of the foundation and the walls of the house. If topography indicates adjacent areas of rock and soil you may have to adjust the position of the house within the lot, or raise or lower the elevation of the foundations to make sure the house is built either all on soil or all on rocks.

If a big rock is in the way of the foundations or a sewer trench, good engineering practice dictates that the rock be blasted to about 10 inches below the bottom of the footing or the sewer pipe. These 10 inches should be filled with clean sand to eliminate hard spots that can cause cracking or breaking of the foundations or sewer pipes. In addition, the footings in the vicinity of the blasted rock should be reinforced by steel bars. The amount and extent of the reinforcing bars should be determined by a structural engineer.

It is also a good practice to reduce the noise levels of blasting by placing the explosives deep inside the holes. This may mean increasing the number of the blasting holes or their size. Another noise control measure is to limit blasting as much as possible during cloudy weather. Clouds cause what is known as the roofing effect. It is like blasting indoors.

Do not expect the bottom of the blasted area to be perfectly smooth. The footings will have to be stepped up and down following the surface of the rocks.

Contracting for Blasting

Blasting is dangerous and should be assigned only to a professional with a good reputation and solid experience. You must ask the bidders for references, and check out those references regarding the bidders' performance, character, and price.

You may require that the blaster be bonded, which means that he must post a surety bond of a face value exceeding the value of the contract. By this, the surety company agrees to complete the contract and pay all costs up to the face value of the bond if the contractor defaults.

Blasting is very expensive and prices can vary substantially from one blaster to another. Therefore, it is in your interest to solicit fixed-price bids from several contractors. Each bidder should be given a set of the foundation drawings with the depth of blasting clearly indicated.

Your contract with the blaster should be in writing. It should include the following specifications in addition to those given in Chapter 3:

1. The blaster shall be bonded and carry all required insurance to cover this particular job.

2. The blaster shall obtain the required permits and licenses from the authorities, and shall comply with all the rules and regulations of the federal, state and local governments.

3. The blaster shall perform the necessary seismographic analysis to determine the vibration characteristics of the site.

4. The work shall be performed in accordance with the best acceptable standards of the trade.

5. The blaster is responsible for the lives of his workers and passers by and the lives and properties of the neighbors, and shall compensate them in the case of injury resulting from his work without any obligation, expense, or responsibility to the owner.

6. Blasting shall be in accordance with the drawings provided by the owner.

7. The work shall be completed by —— .

8. All the above work shall be done for the lump sum of ——dollars.

Every page of the contract should be signed by both yourself (the owner) and the blaster (the contractor).

After signing the contract, the blaster gets a permit from the local authorities, and the required insurance policy and bond. He proceeds with blasting in accordance with the foundation drawings until blasting is completed. Afterward, you should check that the excavated area is wide enough to accommodate the foundations. It is a good policy to hold a part of the blaster's payment until the mason lays the foundations and confirms that blasting was done in accordance with the drawings.

CHAPTER 21

Foundations

The function of the foundation is to safely support the house throughout its life. It also makes up the walls of the basement, if there is one. The load of the house consists of its own weight, including floors, walls, roof, and exterior siding, and live loads such as people, furniture, and snow. Foundations also transfer the horizontal forces resulting from wind and earthquakes to the bearing soil.

Most house foundations consist of **perimeter footings and walls.** The walls transfer the house's loads plus their own weight to the footings; the footings distribute the loads of the house and foundation walls onto the bearing soil. Intermediate house loads are supported by isolated footings. Decks and porches are usually supported by pier foundations.

Footings

The footings are the base on which the foundation walls rest. They must be strong and rigid enough to sustain the house loads and distribute them onto the **bearing soil** without breaking or deforming. The most common material for construction of footings is **ordinary concrete,** which means concrete without reinforcing steel. If the bearing soil is not uniform, the concrete should be reinforced with steel bars (**reinforced concrete**).

Ordinary Concrete Ordinary concrete is a mix of coarse aggregates, which may be gravel or crushed stones, fine aggregates (sand), Portland cement, and water. To produce strong and dense concrete, the mix should be such that the fine aggregates fill the voids in the coarse aggregates, and the cement fills the voids in the fine aggregates. The exact proportion of a mix depends on the natural grading of the coarse and fine aggregates and on the desired strength of the concrete. A strong mix consists of three cubic feet of coarse aggregates, two cubic feet of sand, one bag (94 pounds) of cement, and six gallons of water. If the aggregates are wet (as after a rainfall), the amount of the mixing water must be reduced by an amount equal to that in the aggregates.

Wet concrete should be neither too

dry nor too fluid. Dry concrete is difficult to mold and may **honeycomb,** meaning it will form cavities. Fluid concrete results from too much water, reducing the strength of the concrete significantly. Cement starts to **set** (harden), about one hour after it is mixed with water. Thus, concrete that has been mixed with water more than one and a half hours earlier must not be used for construction.

Concrete should not be poured if the temperature is below 45° Fahrenheit. It is essential that the temperature does not go below 45°F during the following three days. It is also not advisable to pour concrete if the temperature goes above 90°F.

Ready-Mix Concrete   For efficiency and reliability, buy **ready-mix concrete** from a concrete plant. First, you have to be sure that the construction site is accessible to the huge concrete delivery truck. A week before the day of pouring, either you or the contractor—whoever is in charge—should call the plant and give the dispatcher the amount of concrete in cubic yards (this amount should be shown in the bill of materials, or else calculated by the mason or yourself), the strength of the concrete, and the exact date and hour of pouring. The amount ordered should be at least one cubic yard more than the calculated quantity to compensate for the droppings. It is better to have some left over than to run out.

The concrete will be delivered to you in a truck with a rotating drum. In most cases, the coarse aggregates, sand, and cement are loaded in the drum, and the water in a separate tank. The drum rotates to mix the aggregates and cement

while the truck is on its way. The water is added to the dry mix only when the site is ready for pouring. This eliminates the possibility of the concrete setting in the drum if the truck arrives with wet concrete before the site is ready for it.

Most houses are built with 2000 pounds per-square-inch (**PSI**) concrete, which means that the concrete will have a crushing strength of at least 2000 PSI when tested after 28 days of pouring. My recommendation is to use 3000 PSI concrete, instead. It is 50 percent stronger than the 2000 PSI concrete, yet costs only 5 percent more. The 3000 PSI concrete has more cement in it, and looks denser and darker (greener) than the 2000 PSI concrete. Ask the plant delivering the concrete for a certificate indicating its strength.

Reinforced Concrete   If the bearing soil is weak or nonuniform, it will settle unevenly under the weight of the house, causing the ordinary concrete footing to break. This is because ordinary concrete is weak in resisting tension. To compensate for this deficiency, reinforcing steel bars, which are very strong in resisting tension, are embedded in the concrete. Figure 21.1 illustrates typical footing reinforcements.

Footing Forms   Fresh concrete is semifluid. Therefore, it has to be poured in **forms** until it hardens, then the forms are stripped (removed). The exact position of the forms is determined by the nails in the batter boards established by the land surveyor (Chapter 17). Most forms are made of wood. They are secured in place by

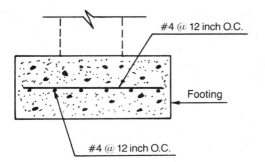

Figure 21.1 Reinforcement of footing to withstand weak or nonuniform soil.

means of stakes driven into the ground along their outer sides. Opposite forms should be tied together every two or three feet to prevent them from bulging during concrete pouring. The top edges of the forms must be perfectly horizontal. This is accomplished by using a transit, level, theodolite, or by stretching lines between the nails of opposing batter boards. The face of the forms in contact with concrete is usually oiled to preserve the forms and facilitate their stripping.

Instead of building forms, footings may be poured in dugout trenches with the sides of the trenches acting as

forms (see Chapter 19). Again, this method is not recommended if subsurface water exists, because it does not allow the drainpipe to be constructed at the bottom of the footing.

Size of Footings and Constructional Details The size of the footings should be as shown on the design drawings or blueprints. If the natural grade is sloped, footings must be poured stepped rather than sloped in order to prevent their sliding. Figure 21.2 illustrates the correct and incorrect methods of constructing footings in a sloped lot. In no place should the base of the footing be above the frost line.

If an area that is in the way of the footings is overexcavated, either accidentally or to remove bad soil, it should be refilled with either clean sand, gravel, crushed stones or concrete, but never with loose fill.

For better bonding between the footings and the walls above them, and to increase the walls' resistance to the lat-

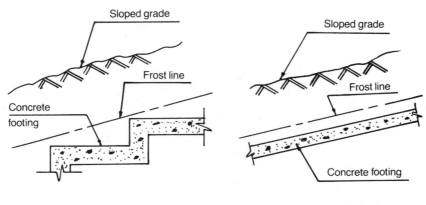

Correct Method   Incorrect Method

Figure 21.2 The correct and incorrect methods of constructing footing on sloped grades.

eral forces of the backfill, install **dowels** consisting of #4 ($\frac{1}{2}$-inch diameter) reinforcing steel bars—called rebars—at 24 inches o.c. (Figure 21.3).

If the foundation walls are to be constructed of poured concrete, the top surface of the footing should have a 2 × 4-inch key along its centerline. This key has two important functions: (1) to secure the foundation walls in place, and (2) to provide more resistance to subsurface water seepage into the basement, because water will have to travel a longer distance (Figure 21.3).

Before pouring the footings, the town inspector must be notified to verify that the forms are in the correct positions and the soil upon which the footings will be poured can support the weight of the house.

Pouring the Footings Any water within the forms must be drained as explained on page 169 before pouring the footings. Once started, concrete should be poured continuously until

the footings are completed. Immediately after pouring, the concrete should be compacted by means of a vibrator or by poking the concrete, particularly at the corners, with steel rods or 1 × 1-inch wood pieces. Concrete should not be allowed to drop freely more than five feet, otherwise it will segregate with the coarse aggregate dropping faster than the sand and cement mortar. The freshly poured concrete should be protected from the rain by covering it with plastic sheets until it hardens.

Concrete Curing The morning after pouring, concrete should be watered generously, a process called **curing.** It enhances the *hydration of the cement,* which significantly increases the strength and water tightness of the concrete. Ideally, concrete should be cured for seven days or until it is backfilled, but the first morning after pouring is the most crucial. Curing is particularly important in hot weather

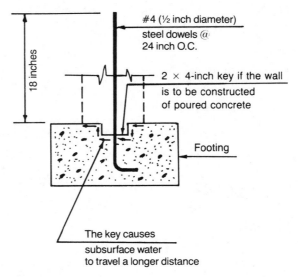

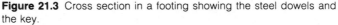

Figure 21.3 Cross section in a footing showing the steel dowels and the key.

because the moisture in the concrete evaporates quickly and this prevents the hydration of the cement.

After the footings are left to harden for a few days, construction of the foundation walls may begin. Walls are constructed of either poured concrete or concrete blocks.

Poured Concrete Walls

Poured concrete walls are more expensive than concrete block walls, but they are stronger and resist water seepage better. Their high initial cost is offset by the higher resale value of the house.

Wall Thicknesses Wall thicknesses should be as indicated in the design drawings, but in no case less than 8 inches. The thickness may be increased to 10 inches for walls supporting brick

or stone veneer exterior siding. In this case, the top of the wall is set back by 4 inches (Figure 21.4a) to create a **shelf** upon which the bricks or stones rest. Another alternative for supporting the brick or stone veneer is to keep the foundation wall thickness at 8 inches and build a 6-inch thick concrete block wall adjacent to it starting from the top of the footings up to the grade elevation (Figure 21.4b).

Formwork The construction of the walls starts with installing the **formwork.** The face of the outer forms is laid along the building lines established by the land surveyor. It is very important that the forms be laid perfectly straight and plumb, meaning perfectly vertical. Furthermore, the carpenter who installs the formwork must ensure that the house will be square by measuring the diagonal lines between opposite corners of the forms and seeing that they are equal in

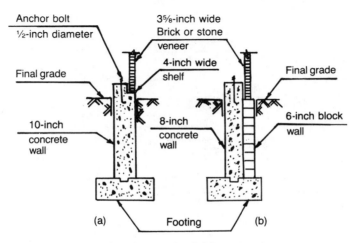

Figure 21.4 Methods of supporting brick or stone veneer.

length. The formwork must be braced adequately to prevent it from tilting or swaying during concrete pouring. The top of the walls should be marked on the forms using a transit or level.

Most forms consist of standard sections that can be assembled and disassembled with relative ease. It is an inexpensive method of construction, but it also does not produce smooth concrete surfaces. For first-class concrete work, you may buy the timber and hire a competent carpenter to install the forms, and an experienced mason to pour the concrete. This method is expensive, but it produces high quality concrete with smooth surfaces that need not—and should not—be painted.

Inserts   All inserts must be placed in the forms before pouring the concrete. The most important inserts are:

Windows. Inserts for windows consist of wood boxes called **bucks.** Their outer dimensions are equal to those of the window openings, and their width is equal to the distance between the faces of the forms. These boxes are placed inside the forms at the exact locations of the window openings. They are held in place by nailing them to the forms. The wall above the boxes should be reinforced by 2 #4 bars (Figure 21.5).

Doorways. Inserts for doorways are similar to those of windows.

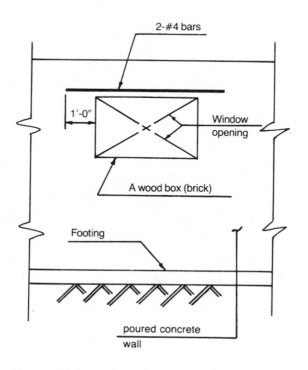

Figure 21.5 Inserts for a window opening.

Anchor Bolts. Anchor bolts are used to tie the wood frame of the house to the foundation walls. They are made of ½-inch diameter steel rods with one end threaded and the other bent or hooked. The length of the embedded part is usually 12 inches. Anchor bolts are embedded in the top of the walls (Figure 21.4) every four feet.

Flue Pipes. Flue pipes for the boiler's exhaust gases and for the ash pit of the fireplaces must be placed in the forms in accordance with the design drawings.

Blocks for Brick Veneer Shelves. If the exterior siding is brick or stone veneer built on shelves, wood blocks should be inserted in the forms along the top of the walls to create these shelves.

For tightness, foundation walls must not have any holes or openings below the final grade elevation.

Pouring the Walls   Before pouring begins, the surface of the footing that will be in contact with the new concrete should be cleaned of flakes, dirt, grease, oil, or any other impurity that may prevent the new concrete from bonding to the old one. In addition, the old surface should be moistened. If this is not done, the old concrete absorbs the moisture from the fresh concrete causing it to crack along the line of contact between the two surfaces.

Walls should be poured continuously in layers no more than 12 inches high until the walls are completed. If this is not feasible for any reason, the stoppage or break should be at a diagonal. The surface of the break should be roughed and keyed before the concrete sets. For better bond between old and new concrete, it is recommended to insert dowels made of #4 bars every two feet along the breakline. The dowels should be about two feet long with one foot embedded in the old concrete. Before pouring is resumed, the old surface should be treated as explained previously. Immediately after pouring, concrete should be thoroughly compacted using gas or electric vibrators.

Concrete and Cinder Block Walls

Concrete and cinder blocks are widely used in constructing foundation walls. They are cheaper, lighter and easier to work with than poured concrete. However, they are weak in resisting lateral forces and water seepage.

The most common widths of concrete and cinder block walls are 8, 10, and 12 inches. Some building codes require that block foundation walls should not be less than 12 inches wide; others specify that the first five courses should be at least 12 inches wide, and the rest may be 8 inches. (Check the local building code.)

Shapes and Sizes of Concrete and Cinder Blocks   There are more than 30 different shapes and sizes of blocks. Each one is suited to a particular application. The basic unit is called the stretcher. It is 16 inches long, 8 inches high and has three vertical holes in it. Other shapes include corner, double corner, partition, jamb, and beam. Building material suppliers will be more than happy to show you their inventory, once they know that you are building a house.

Block Wall Construction   The first step in constructing block walls is to mark the building lines on top of the footing by stretching lines between the nails in the batter boards that were established by the land surveyor. After marking all the lines, the mason must ensure that the house will be square by measuring the diagonals between opposite corners of the walls and seeing that they are equal in length.

The blocks are bonded together with mortar consisting of sand, Portland cement, and water. Most mortar mixes include some lime to increase their workability and reduce shrinkage cracking caused by cement hydration. Building material suppliers carry bags of dry ready-mix mortar with and without lime.

Concrete block walls can take vertical loads satisfactorily, but they have little resistance to lateral loads caused by the backfill. Therefore, they have to be braced until the ceiling of the basement is constructed. Thereafter, the joists of the ceiling provide the walls with lateral support.

The top of the block wall must be solid in order to provide a rigid base for the wood framing of the house and to accommodate the anchor bolts. This can be accomplished by either filling the holes of the top two block courses with mortar or by capping the wall with beam type blocks filled with reinforced concrete.

Block Wall Reinforcing   It is recommended that block walls be reinforced both vertically and horizontally to increase their resistance to lateral forces and cracking. The ceiling joists provide the walls with lateral support at their top, but this does not ensure the walls' stability at their midheight.

Adequate vertical reinforcement

consists of #4 ($\frac{1}{2}$-inch diameter) steel bars placed in the holes of the blocks and distributed as follows: one bar every two feet along the entire length of the walls, and one bar along each side of door and window openings. The bars should extend over the entire height of the walls. The holes around the bars must be filled with mortar or grout.

Horizontal reinforcement may consist of galvanized wire mesh or two $\frac{1}{4}$-inch diameter reinforcing bars installed along the entire length of the walls every second course. Intersecting walls are tied together by means of steel ties.

Slab Foundations

Slab foundations are simple and inexpensive, but they have limitations. First, they do not accommodate basements. Second, they cannot support more than one story. Slab foundations consist of two parts: a perimeter concrete wall or footing, and a poured on grade concrete slab. All inserts such as water and sewage pipes, anchor bolts etc., must be in place before pouring the concrete.

The construction of slab foundations in warm areas is different from that in cold areas as explained in the following:

Warm-Area Construction   The top soil is excavated and a perimeter trench one foot wide and about two feet deep is dug. The concrete of the perimeter wall or footing is poured in the trench without forms. The concrete slab is poured over a moisture (vapor) barrier

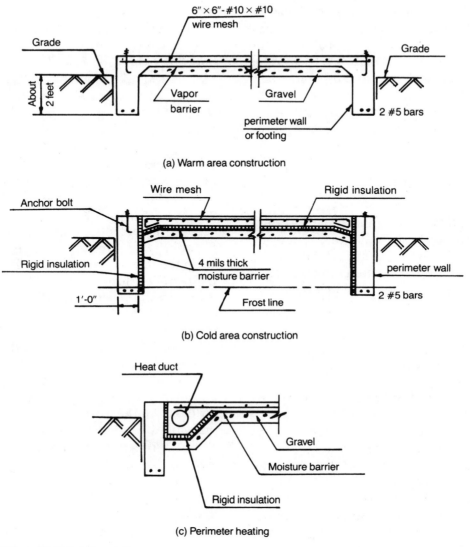

(a) Warm area construction

(b) Cold area construction

(c) Perimeter heating

Figure 21.6 Slab foundation.

consisting of 4-mil thick polyethylene plastic sheets underlain by four inches of gravel (Figure 21.6a). The slab is usually four inches thick. It is recommended that the bottom of the wall be reinforced by two #5 ($\frac{5}{8}$ inch diameter) bars, and the slab be reinforced by at least a $6 \times 6 - \#10 \times \#10$ wire mesh (see Figure 21.6a).

Cold-Area Construction Construction of slab foundations in cold areas is different from that in warm areas in two respects: (1) the bottom of the perimeter wall must be below the frost line, and (2) the walls and slabs should be thermally insulated (Figure 21.6b). Because of the thermal insulation, the concrete slab is not cast monolithically with the perimeter wall, as for warm area construction, and this weakens the slab. To offset this weakness, the thickness of the slab at the edges should be increased and the wire mesh bent (Figure 21.6b). Alternatively, the slab may be heated by an embedded heat duct (Figure 21.6c).

Isolated Footings

Isolated footings are used to support intermediate weights of the house within the basement area. They are poured below the concrete slab of the basement. Most isolated footings are 24 inches square and 12 inches high.

Piers

Piers are used to support small outdoor loads such as decks and porches. Piers look like and function as short columns. Their cross section may be square or circular. Their construction begins with digging a square or circular hole to below the frost line or until firm soil is reached. Next, a footing is poured in the bottom of the hole. On top of the footing, a square or circular pier is constructed to the height called for in the design. Square piers may be constructed of concrete blocks, bricks or stones. Circular piers are usually constructed of concrete poured in forms made of cardboard tubes. After the concrete hardens, the tubes are peeled off.

Proposed Design for Earthquake-Prone Areas

On October 1, 1987, a powerful earthquake shook the Los Angeles area kill-ing at least six people and injuring more than a hundred. It caused extensive damage to property and set many buildings on fire. Scientists say that this earthquake, which registered 6.1 on the **Richter Scale,** is nothing compared to the big one due to occur within the next 50 years along the **San Andreas** fault that will register 8.0 on the Richter Scale. (Each unit increase in this scale corresponds to a thirtyfold increase in the energy released by the quake.)

All this prompted me to develop a foundation design that is much stronger than what is required by the **Uniform Building Code.** This design is empirical, because we do not know the exact forces that an earthquake of magnitude 8.0 will impose on buildings. It is correct to state that the damage to a house will be inversely proportional to the strength of its foundations. In other words, if you are in an earthquake-prone area, you will probably sleep better knowing that the strength of your foundations is substantial. Fortunately, the extra cost involved in this design is only a small percentage of the total cost of the house.

The basic premise of this design is that in order to resist the seismic forces of a powerful earthquake, the foundations must be both strong and flexible. Thus, they must be made of reinforced concrete. To attain both strength and flexibility, the reinforcing steel should be heavy, but the concrete must not be thick. The heavy steel will carry most of the stresses caused by the earthquake's cyclic forces, while the fact that the concrete is not too thick will give the foundations flexibility to absorb some of the quake's energy.

Under the initial shock, the tension

forces resulting from the earthquake will cause the concrete of one side of the foundations to crack. The reinforcing steel at this side will immediately carry all the tension forces. When the initial forces are reversed—a characteristic of earthquakes—the concrete side that cracked goes back to its original condition, and the opposite side cracks, causing the steel bars of that side to resist all the tension forces. In a typical earthquake, this loading cycle is repeated many times in a very short period of time. Steel can resist reversed loading satisfactorily, but ordinary concrete cannot. Therefore, the footing and walls must have double reinforcement, meaning an equal amount of steel on each side. The proposed design is shown in Figure 21.7.

Waterproofing Foundation Walls

The under-grade parts of the foundation walls must be **waterproofed** to protect them from the moisture and humidity of the surrounding soil. Waterproofing also acts as a first line of defense in preventing subsurface water from seeping into the basement. As mentioned earlier, the most effective line of defense against water seepage is good quality concrete.

Before waterproofing concrete block walls, they must be coated with a $\frac{3}{8}$-inch thick layer of **stucco,** then left for a few days to dry. Poured concrete walls do not need coating. If there is no subsurface water, the walls are treated with one coat of **trowelled-on asphalt.** Where subsurface water is

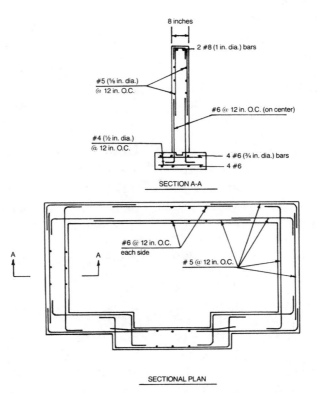

Figure 21.7 Proposed foundation design for earthquake-prone areas.

moderate, the walls are treated with one coat of trowelled-on asphalt, followed by one layer of 15-pound **asphalt-saturated felt** with a minimum lap of 6 inches, then another coat of trowelled-on asphalt to seal the felt. If subsurface water is serious, the walls should be waterproofed as just explained and in addition, a drainpipe is installed. The seriousness of the subsurface water problem can be determined by either the excavation contractor, a geotechnical or structural engineer.

Drainpipe

Where subsurface water is high, a **drainpipe** must be installed all around the footings of the house. Its function is to drain the subsurface water continuously, and this keeps the basement dry. The drained water is discharged into the sewer line, or dry well.

The drainpipe may consist of a 4-inch diameter **perforated plastic pipe**. The subsurface water enters the pipe through its holes. To prevent the surrounding soil from clogging the holes, the pipe should be laid on three inches of gravel or crushed stones, then covered all around with six inches of the same material (Figure 21.8). Instead of perforated pipe, **clay pipes** may be laid end to end. The pipes must not be cemented or glued together, in order to allow the subsurface water to seep into the pipes through the joints. Similar to the perforated pipe installation, clay pipes must be laid on three inches and be surrounded by six inches of gravel or crushed stones to prevent the surrounding soil from clogging the joints.

The drained water moves by gravity. Thus, the drainpipe must be sloped

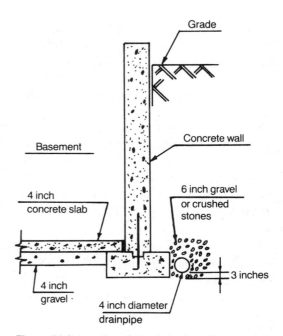

Figure 21.8 Location of the drainpipe with respect to the footing.

continuously toward the point of discharge. The slope varies between $\frac{1}{4}$ and $\frac{1}{2}$ inch per foot. The highest point of the drainpipe must be a few inches lower than the floor of the basement.

Underground Sewage and Water Pipes

As soon as the foundations are completed, the plumber should be called in to install all underground sewage and water pipes before pouring the concrete slab and backfilling.

Foundation Insulation

Foundation insulation consists of **rigid boards** in order to resist the pressure and humidity of the backfill and the loads of the concrete slab. The minimum value of the thermal resistance of the insulation, **R**, is determined by the local building code. Insulation having R = 11 (see Chapter 31) is adequate for most areas. (The value of R should be clearly printed on the face of the boards.)

Here are the most widely used methods of foundation insulation:

1. Rigid boards are installed against the inner face of the foundation wall starting from below the frost line up to the top of the concrete slab (Figure 21.9a).

2. Rigid boards are installed underneath and around the edges of the concrete slab (Figure 21.9b).

3. Rigid boards are installed against the outer face of the foundation walls starting from below the frost line up to the top of the concrete wall (Figure 21.9c). It is very important that the soil around this insulation be well drained.

Slab foundation insulation should be as shown in Figure 21.9.

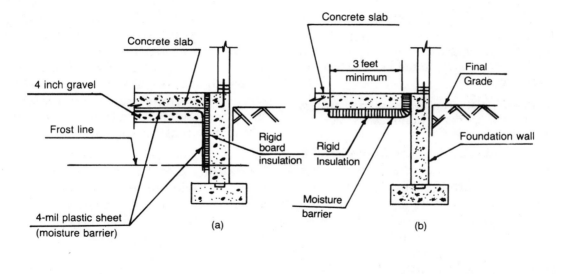

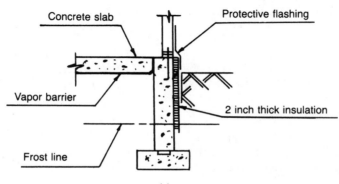

Figure 21.9 Foundation insulation.

Crawl Spaces

Some homeowners prefer to build their houses with a crawl space instead of a basement to reduce construction costs. There are several differences between crawl spaces and basements:

1. The minimum clear height of crawl spaces is only 18 inches (Figure 21.10), as compared with 7'-6" for basements.

2. The floor of crawl spaces is fill, as opposed to poured concrete slab for basements.

3. Crawl spaces are usually unheated.

4. Crawl spaces do not require drainpipes.

5. Generally, if crawl space foundation walls are constructed of concrete or cinder blocks, they do not need to be reinforced.

6. Intermediate support for inner girders of crawl spaces is provided by piers.

The engineering rules for constructing the foundations of a house with a crawl space are the same as those for a house with a basement. Briefly, the footings should be poured on undisturbed soil, never on fill. If the soil is nonhomogeneous, the footings should be reinforced as shown in Figure 21.1. In cold areas, the bottom of the footings must be below the frost line.

Termites

Termites are insects that attack wood. There are two groups of termites: **subterranean** and **dry-wood.***

Subterranean Termites Subterranean termites are common throughout most

of the United States where they cause considerable damage in some areas. Termites thrive in moist, warm soil containing wood or other cellulosic material. They get to the wood of the house by building a tube of mud over foundation walls or in wall cracks, or on supports leading from the soil to the house. These tubes vary from $\frac{1}{4}$ to $\frac{1}{2}$ inch across. Subterranean termites eat the interior of the wood, and may cause much damage before they are discovered.

Dry-Wood Termites Dry-wood termites are found in the South from Florida to California. They attack all types of wood including furniture. They cut across the grains of the wood and excavate chambers connected by tunnels. Dry-wood termites remain hidden in the wood and are seldom seen, except when they make dispersal flights.

Protection against Termites The best time to provide protection against termites is during construction of the foundations of the house. The following precautions are important:

1. Remove all woody debris from the soil at the building site.

2. Keep the soil under the house as dry as possible.

3. All cracks in the foundation walls must be patched. Termites are known to crawl through hidden wall cracks to reach the wood of the house. In this regard, foundation walls made of poured concrete are superior because they crack less than walls made of concrete or cinder blocks.

4. No part of the wood framing should be in contact with the soil.

5. Build the sill plate (which is laid on top

---

*U.S. Department of Agriculture, Agriculture Handbook No. 73.

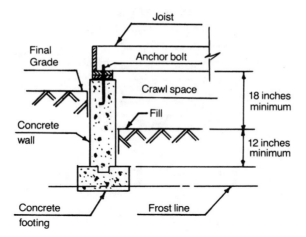

Figure 21.10 Partial cross section in a crawl space.

of the foundation wall) of **pressure-treated** wood, because it is immune to termite attack.

6. If subterranean termites are known to be in the area, the soil near the foundations or under the entire slab foundation should be treated with pesticides (by a professional exterminator).

7. Inspect the crawl space area annually for termites.

8. Install a continuous metal shield on top of the foundation wall.

Backfilling

After the waterproofing is finished, and the drainpipe, underground sewage and water pipes and foundation insulation are installed, backfilling begins. The machines used for backfilling are the same as those used for excavation. The machine operator must be experienced. An incompetent operator may damage the foundation walls, causing them to crack or break. As a matter of precaution, the machine should not get closer than two feet from the wall. This area should be backfilled manually.

Contracting for the Foundations

Poured concrete footings and walls may be contracted at a fixed price or per cubic yard of poured concrete. The price should include labor and material.

Concrete and cinder block walls may be contracted at a fixed price or per square foot. Two block layers and one laborer (who mixes the mortar and delivers the mortar and blocks to the block layers) can build 200 square feet of 8-inch thick blocks per day, on the average.

Waterproofing should be contracted at a fixed price based on the drawings and specifications. A mason can install 250 square feet of stucco per day. A waterproofer can trowel on 400 square feet of asphalt or install over 300 square feet of 15-pound asphalt saturated felt per day.

The drainpipe should be contracted on a lump sum basis. The same goes for thermal insulation. Backfilling may be assigned to the excavation contractor, or you may rent a machine and operator by the day.

CHAPTER 22

House Frame

The house frame, also called rough framing or rough carpentry, is the supporting structure of the house. It is to the house what a skeleton is to the human body. Like the skeleton, it consists of many parts held together: sills, studs, joists, girders, rafters, sole plates, top plates, subflooring, sheathing, etc. Most of the house frame is constructed of softwood lumber. But some parts, such as long girders, columns and posts, may be made of steel.

House framing is constructed by a rough carpenter. Before starting construction, the carpenter must check that the foundation perimeter walls are straight and square and that their top is perfectly horizontal. The carpenter has some room to adjust for minor irregularities since the width of the wood framing is usually less than the width of the foundation walls. If the top of the foundation wall is not perfectly horizontal, a mason may have to be called in to level it with a layer of mortar.

There are three methods of house framing: (1) **platform framing,** (2) **balloon framing,** and (3) **post and beam**

framing. Before explaining these methods, the basic characteristics of lumber and steel are reviewed. This information is important particularly if you intend to buy the materials and contract for labor only.

Characteristics of Lumber

The basic characteristics of lumber are:

Nominal and Standard Dressed Sizes Lumber has two sizes: **nominal** and **standard dressed** or **dressed.** The nominal are the sizes used in the design drawings and to price the lumber; the standard dressed are the actual sizes. For example, a 2 × 4-inch piece of lumber is actually $1\frac{1}{2} \times 3\frac{1}{2}$ inches. Table 22.1 shows the nominal and the corresponding standard dressed sizes of the lumber most commonly used in house framing.

Lumber Lengths Lumber is sold in even lengths, for instance, 6, 8, 10, and 12 feet. Architects and building

Table 22.1 The Nominal and Standard Dressed Sizes of Lumber Most Commonly Used in House Framing

Nominal (inches)	Standard Dressed (inches)	Nominal (inches)	Standard Dressed (inches)
1×3	$\frac{3}{4} \times 2\frac{1}{2}$	2×4	$1\frac{1}{2} \times 3\frac{1}{2}$
1×4	$\frac{3}{4} \times 3\frac{1}{2}$	2×6	$1\frac{1}{2} \times 5\frac{1}{2}$
1×6	$\frac{3}{4} \times 5\frac{1}{2}$	2×8	$1\frac{1}{2} \times 7\frac{1}{4}$
1×8	$\frac{3}{4} \times 7\frac{1}{4}$	2×10	$1\frac{1}{2} \times 9\frac{1}{4}$
1×10	$\frac{3}{4} \times 9\frac{1}{4}$	2×12	$1\frac{1}{2} \times 11\frac{1}{4}$
1×12	$\frac{3}{4} \times 11\frac{1}{4}$	2×14	$1\frac{1}{2} \times 13\frac{1}{4}$
4×4	$3\frac{1}{2} \times 3\frac{1}{2}$	6×8	$5\frac{1}{2} \times 7\frac{1}{2}$
4×6	$3\frac{1}{2} \times 5\frac{1}{2}$	8×8	$7\frac{1}{2} \times 7\frac{1}{2}$

designers size the rooms of the house to accommodate these standard lengths. Thus, if you desire to enlarge a room in the direction parallel to the joists, the extension has to be in two-foot multiples; otherwise the carpenter will have to cut the lumber, thus wasting material and labor time. In platform framing, studs can be bought in a precut length of 7'-9" to produce a clear ceiling height of 8'-0".

Green and Dry Lumber   **Green lumber** has a moisture content of more than 19 percent. **Dry lumber** has a moisture content of 19 percent or less. Lumber used in construction has about 15 percent moisture content.

Lumber Species   Several lumber species are used in house construction: eastern and western hemlocks, and western white pine are of lower quality than Douglas fir and southern yellow pine. The higher quality species are recommended, particularly since the difference in price is not substantial. Most lumber yards do not carry all the species mentioned above. Therefore, you have to shop around to get the kind you want.

Lumber Defects   Virtually every piece of lumber has one or more of the following defects:

Knots. Knots are the most common defect. A knot affects the strength of a piece of lumber in that it interrupts the path of the fibers. The seriousness of a knot depends on its size and on the structural function of the lumber. For example, knots have little effect on lumber that is under compression, such as studs.

Check. A check is a break across the annual rings of the wood. All girders, joists and studs should be free of checks.

Shake. A shake is a break in between the annual rings of the wood. All girders, joists and studs should be free of shakes.

Warp. Warp results from uneven drying of wood. Pieces of lumber with noticeable warping should not be used in construction.

Wane. A wane occurs when a piece of lumber does not have the full cross section throughout its entire length. The deficient edge is often covered with bark.

Lumber Grades Lumber is graded according to the number, size, and seriousness of the defects in each piece. Several associations have different grading systems. In one system, the gradings of lumber used in framing are: **Construction, Standard,** and **Utility.** Another system uses numbers: **No. 1, No. 2, No. 3** and **No. 4.** In one system, some lumber species are graded as "No. 1 and Better," or "No. 2 and Better." Each piece of lumber is stamped with its grade. Lumber yards sell lumber in mixed grades. For example, if you buy Douglas fir, the yard sells you a mix of "No. 1 and Better" and "No. 2 and Better." The percentage of the mix should be stated in the purchase agreement or contract.

Lumber Shrinkage and Swelling Moisture loss or gain causes lumber to shrink or swell significantly. The ancient Egyptians used this principle to split granite and other hard rock. They drilled a line of holes along the surface they wanted to split and inserted wedges of dry wood in the holes. Next, they poured water on the dry wedges causing them to swell and split the rock.

The amount of shrinkage and swelling varies with respect to the direction of the grain. It is a few percentage points in the direction perpendicular to the grain, and negligible in the direction parallel to the grain.

Lumber Pricing Lumber is sold in either **Hundred Board Feet (CBF)** or **Thousand Board Feet (MBF).** The amount of board feet in a piece of lumber is calculated by multiplying its nominal thickness in inches by its nominal width in feet by its length in feet. As an example, a 2 × 10-inch and 12-foot long piece of lumber has 20 board feet in it ($2 \times (10/12) \times 12 = 20$).

Plywood Plywood is fabricated of thin plies or layers of wood glued together. The grain of each ply runs perpendicular to the grain of the adjoining plies. This construction makes plywood strong in both directions and significantly reduces its tendency to shrink and swell. Plywood is sold in standard 4 × 8-foot panels.

Plywood is graded according to the number and significance of defects in each panel. Starting with the best, the grades are: **A, B, C,** and **D.** The face and back plies may not be of the same grade. For example, **grade C–D** plywood means that the face ply is grade C and the back ply is grade D.

Plywood is also graded in **groups 1 to 5** according to the strength of the lumber of which the face plies are made. Inner plies may be made of lower quality wood. For example, the face plies may be made of group 2, while the inner ply or plies are made of group 4 lumber.

Engineering-grade plywood panels that are used for roof sheathing and subflooring are marked by two numbers with a slash in between, for example, 32/16. These numbers indicate the maximum spacing in inches of the roof rafters and floor joists, respectively, for which the panel is designed. If the second number is zero, for instance, 16/0, it means that the panel should not be used for subflooring.

Plywood panels are also marked either **interior** or **exterior.** The difference is that the exterior type is glued with a waterproof adhesive.

Particleboard Particleboard is made of flakes of wood glued together under pressure and heat. It is sold in 4 × 8-foot panels of various thicknesses. Particleboard is gaining popularity as wall sheathing material because it is cheaper than plywood. However, it has many disadvantages: (1) it is hard to nail or saw, (2) it chips easily, particularly at the edges, and (3) it is heavy.

Characteristics of Steel

Steel has two main advantages over lumber: (1) it is stronger, and (2) it does not shrink or swell with the change of the ambient humidity. (Shrinkage or swelling of girders causes the walls they support to crack.) In spite of these advantages, the use of steel in house framing is limited because it is more expensive than lumber and because steel members are heavy and require a winch or crane to install them.

The principal applications of steel in house framing are:

1. Girders above wide basements. They can span long distances with few or no intermediate supports. This gives the architect or homeowner flexibility in arranging the space of the basement.

2. Girders over two-car garage door openings.

3. Posts and columns for supporting either steel or built-up wood girders over the basement.

Steel girders must be braced laterally because their flanges are narrow. This can be accomplished by fastening the top flange of the girders to the wood joists they support. To control deflection, the depth of a steel girder must not be less than 1/20 of its span.

Platform Framing

Platform framing, also called **western framing,** is the most popular method of house framing nowadays:

1. The sill above the foundation wall consists of a sill and a header, a combination called the **box sill** (Figure 22.1).

2. The frame is built one story at a time. The studs of each story end at the bottom of the top plates (Figure 22.2).

3. The floor of each story serves as a platform upon which the next story is built.

The Sill Sealer Construction begins with installing a sill sealer on top of the foundation wall. It may be made of a foamy material or cement mortar. Its function is to make the contact between the sill and the top of the foundation walls tight.

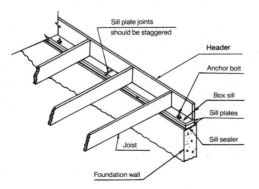

Figure 22.1 Platform framing—the box sill.

The Box Sill　The box sill is a characteristic of platform framing. It has two parts:

1. The **sill,** also called **sill plate,** consists of a 2 × 6-inch plate installed over the sill sealer, then anchored to the foundation wall by anchor bolts. The sill constitutes the base upon which the exterior walls of the house are constructed; you may make it more rigid by installing two plates on top of one another as shown in Figure 22.1. The joints of the two plates should be staggered in order to have at least one continuous plate at any given point.

2. **The Header.** The header is installed on the outer edge of the sill plate(s), as shown in Figure 22.1. It encloses the perimeter around the ends of the floor joists and contributes to supporting the studs. The header should be the same size as the floor joists.

First Floor Framing

The next step after installing the box sill is to frame the first floor. The following description is for framing a floor above a wide basement or a crawl space, since it is the most elaborate.

Intermediate Supports　Intermediate supports are used to uphold the main girders in between the perimeter foundation walls (Figure 22.3). Usually, they consist of steel pipes, lally columns, or wood posts in basements, and masonry piers in crawl spaces. Each steel pipe or lally column has a rectangular base plate welded to each end. The top plate has four holes to accommodate the bolts or lag screws that secure the pipe or column to the girder it supports. The bottom plate is secured in place by the concrete floor slab.

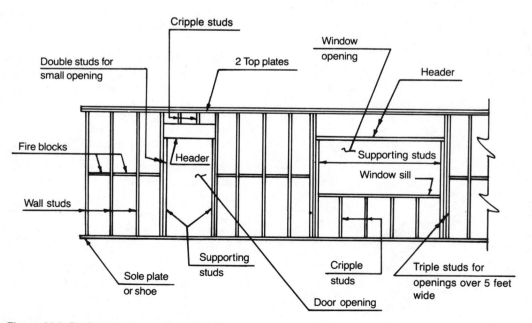

Figure 22.2 Platform framing—a typical wall.

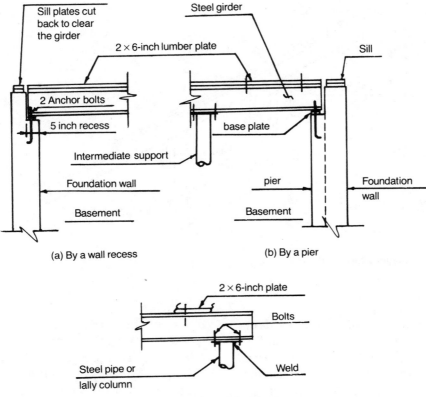

Figure 22.3 Methods of supporting the basement's steel girder.

Wood posts are installed on pedestals or stubs that are a few inches above the finish floor. They should be pinned to their support in a way that does not obstruct the shrinkage or swelling of the wood.

Steel Girder The steel girder may be supported in a recess created in the foundation wall, on a pier, or on steel pipes or lally columns, as shown in Figure 22.3. The steel girder bears on a steel base plate, the function of which is to distribute the girder's heavy loads over a wide area. The masonry on which the base plates rest must be furnished with $\frac{3}{4}$-inch thick grout to create a level surface. The girder is fastened to the masonry by two anchor bolts.

It is more desirable to fasten a 2 × 6-inch wood plate along the top of the upper flange of the girder and use it as a base for the joists.

Built-Up Wood Girders Built-up wood girders consist of three joists nailed together. The joists are two inches

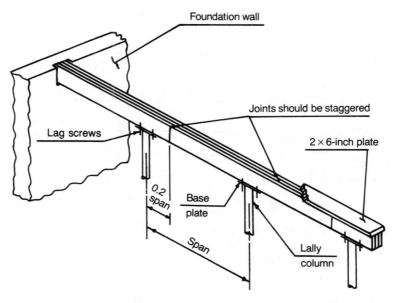

Figure 22.4 A built-up wood girder.

thick and 8, 10, 12 or 14 inches deep. In long girders, the joints should be staggered such that there are at least two continuous joists at any section (Figure 22.4). The joints between the supports should be located at 0.20 of the span (Figure 22.4), because this is where the stress in the girder is the least.

In one method of construction, the joists bear on top of the girder. (In this case, it is a good idea to install a 2 × 6-inch plate on top of the girder as shown in Figure 22.4 to provide a stable surface upon which the joists rest.) The disadvantages of this method are: (1) the girder will shrink while the foundation wall will not, which can cause wall cracking, and (2) it reduces the headroom of the basement.

The shrinkage and headroom problems can be virtually eliminated by installing the joists at the same elevation as the built-up girder. The joists are supported on wood **ledgers** fastened to the girder (Figure 22.5). The disadvantage of this method is that if a ledger fails, it can mean real trouble.

It is not necessary to have the intermediate supports installed before constructing the built-up girders, but the isolated concrete footings should have been poured and allowed to harden. The carpenter can support the girders

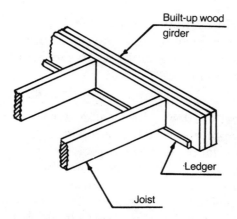

Figure 22.5 The joists are installed at the same elevation as the built-up wood girder.

on temporary wood trusses. Later on, he measures the height of each column exactly. Next, you give the list of column heights to an iron shop for fabrication and delivery. The carpenter installs each column in its proper position, then removes the temporary trusses.

Joists After the sill and the steel or built-up wood girders are installed, the floor joists are laid. The sizes, spacings, and orientation of the joists should be as shown in the design drawings. In most construction, joists are spaced at 16 inches o.c. Few houses have 24-inch joist spacing. Joists should be doubled or tripled under load-bearing walls.

Bridging The purpose of bridging is to brace the free edges of the joists to prevent them from tilting or overturning. Bridging may consist of either 1 × 3-inch diagonal cross bracings or solid blocking (Figure 22.6). The latter is more effective in distributing heavy loads imposed upon a joist to the neighboring joists.

Bridging should be applied at the middle of spans over eight feet long. Blocking is also required between the joists close to their supporting steel girders.

Subflooring Subflooring is the covering of the entire floor. The subfloor consists of wood boards or plywood panels laid on top of the joists and header. It serves several functions: (1) it forms the base on which the finish flooring is installed, (2) it provides the house with rigid bracing in the horizontal direction, and (3) it provides the workers with a working platform.

Board subflooring consists of $\frac{3}{4}$- to 1-inch thick and 6- to 8-inch wide boards. The edges of the boards may be **square** or **tongue-and-groove.** In most construction, the boards are laid diagonally to permit the finish hardwood flooring to be laid either parallel or perpendicular to the direction of the joists.

Plywood subflooring is popular because it is easier to install and has fewer joints than board subflooring, and is strong in both directions. The minimum thickness of the plywood should be as recommended by the manufacturer, but not less than $\frac{1}{2}$ inch. You should increase the thickness to $\frac{5}{8}$ or $\frac{3}{4}$ of an inch if you intend to install marble flooring. Plywood panels should be staggered with $\frac{1}{16}$-inch spacing between the panels and $\frac{1}{8}$-inch spacing at the edges close to the walls to allow for expansion. Panels should be installed with the grain direction of the

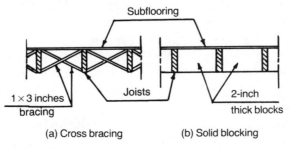

Subflooring

1 × 3 inches bracing   Joists   2-inch thick blocks

(a) Cross bracing   (b) Solid blocking

Figure 22.6 Different types of bridging.

outer ply perpendicular to the direction of the joists.

Some houses are built with ¾-inch thick plywood that serves as both subfloor and underlay for carpeting and resilient finish flooring. This method is not recommended because rain and working conditions are likely to damage the surface of the plywood.

Wall Framing   Wall framing starts after the subflooring is completed. Figure 22.2 shows a typical platform wall frame. It consists of a base made of two-inch thick lumber called a **sole plate** or **shoe,** and a top consisting of two two-inch thick plates called **top plates.** The width of the plates is equal to the width of the studs. The vertical studs are constructed in between the sole plate and the top plates. The studs may be 2 × 4 or 2 × 6 inches in size, spaced mostly at 16 inches o.c.

At window and door openings, the studs should be doubled at the sides of openings less than five feet wide and tripled on the sides of openings of a width of five feet or more. Window openings have a sill at the bottom and a header at the top. Door openings also have a header at the top. Figure 22.2 shows two methods of header construction. The short studs below the window sills or above the headers are called **cripple studs** or **cripples.** In the middle of the exterior wall's height, **fire blocks,** also called **fire stops,** are installed between the studs. They serve two functions: (1) they prevent hot gases and smoke from spreading from one area of the house to another, and (2) they are used as blocking to support horizontally laid wall sheathing.

For ease of construction, it is a common practice to assemble the walls (or parts of them) flat on the subfloor, then raise them into place. After the carpenter makes sure that the walls are aligned and plumb, he braces them temporarily. The sole plate is then nailed to the joists and headers below it through the subflooring. After the ceiling above it is constructed, the temporary braces are removed.

Balloon Framing

Construction of **balloon framing** starts by laying a sill sealer and a sill on top of the foundation walls. It is a good practice to construct the sill of two two-inch thick plates, as in platform framing. The sill is then fastened to the foundation walls by the anchor bolts.

In balloon framing, the studs rest directly on the sill (Figure 22.7). They extend from the sill to the top plate of the highest story. Intermediate floors are supported by 1 × 6-inch lumber called **ribbon** or **ribband,** which is set into the studs (Figure 22.8).

Subflooring ends at the inner edges of the studs, which leaves the spaces between the studs open. These spaces should be completely closed by fire blocks installed at the level of each floor (Figures 22.7 and 22.8). In addition, local building codes may require that fire blocks be installed between the studs at the midheight of each floor. Other details such as steel and built-up wood girders, framing around window and door openings, joist spacings, and subflooring thicknesses should be the same as for platform framing.

From the structural point of view, balloon framing is superior to platform framing in two respects: (1) it is stronger because the studs extend to the full height of the house without joints, and (2) the vertical shrinkage of the frame is negligible since the entire height is

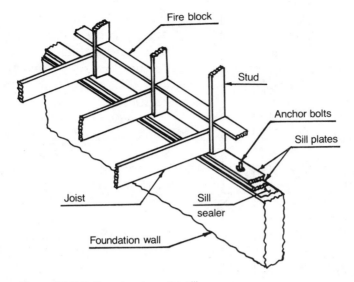

Figure 22.7 Balloon framing—the sill.

made of studs. Nevertheless, this method of framing is not popular because it costs more than platform framing.

Post and Beam Framing

In **post and beam** framing, the floors and ceiling rest on large beams, also called **girts,** that are supported by large

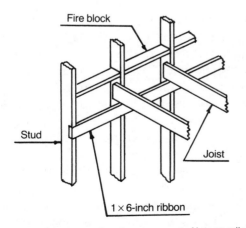

Figure 22.8 Balloon framing—support of intermediate floors.

posts (Figure 22.9). The posts are spaced far apart, which provides flexibility in arranging the interior of the house, and in having wide glazed openings in the exterior walls. Usually, the house has a Cathedral ceiling formed by wood decking. The posts, beams, and decking are exposed and stained for decorative purposes.

Construction of the main frame starts with installing a 4 × 6-inch sill on top of the foundation wall. The sill is fastened securely in its place by anchor bolts. It is to be mentioned that a 4 × 6-inch sill plate is much stronger than two 2 × 6-inch plates. The four corner posts are installed, extending along the entire height of the house. Their size varies between 4 × 6 and 8 × 8 inches, depending on the size and number of stories in the house. Intermediate posts are also installed as called for in the design. The beams are then fastened to the sides of the posts. Usually, beams and posts are of the same size for better appearance. Intermediate bearing walls with 2 × 4-inch studs are used to

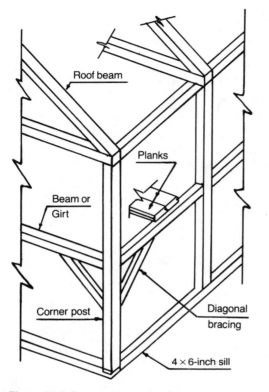

Figure 22.9 Post and beam framing.

partition the floors and, in the meantime, to support parts of the loading. This type of construction can be used for Contemporary style houses only.

The floors and roof decking are formed by two- or three-inch thick planks with tongue-and-groove edges. The planks bear directly on top of the floor and roof beams.

Post and beam construction is weak in resisting horizontal forces due to lack of exterior wall sheathing. If the exterior walls do not have much sheathing, the posts and beams must be provided with heavy diagonal bracing.

Roof Framing

The main components of the **roof framing** are the **ceiling joists, rafters** and **ridge board,** and **collar beams.** In addition, gable roofs have **end studs** at their end gables as shown in Figure 22.10a; hip roofs have **hip, common, and jack rafters** as shown in Figure 22.10b.

The purpose of the ceiling joists is to: (1) support the ceiling finishes, (2) support the floor of heated attics, (3) support the thermal insulation of unheated attics, and (4) resist the horizontal forces of the rafters.

In order to resist the horizontal forces of the rafters, the joists must run parallel to the rafters, and they both must be fastened securely together. If the joists are not laid parallel to the rafters, collar beams or tie beams must be installed to tie opposite rafters together (Figure 22.10a). Another solution is to support the ridge board with vertical posts that rest on bearing walls. If none of these precautions are taken, the rafters will open outward, taking the exterior walls with them.

The rafters are installed after the ceiling joists and ridge board are laid and fastened in place. Their outer ends rest on the top plates. Before lifting them in place, the rafters should be precut to the exact lengths. Their outer end should be cut to the proper angle to provide a horizontal surface for fastening the soffit boards and a vertical surface for fastening the facia boards (see Chapter 23). They should also be notched where they bear on the top plates to create a broad surface of contact. Their inner end should be cut to the proper angle to fit the ridge board tightly.

The ridge board ties opposite rafters together at their upper ends. Therefore, it needs to be rigid. It should be 2 × 10 inches for 2 × 8-inch rafters. In big houses, the recommended

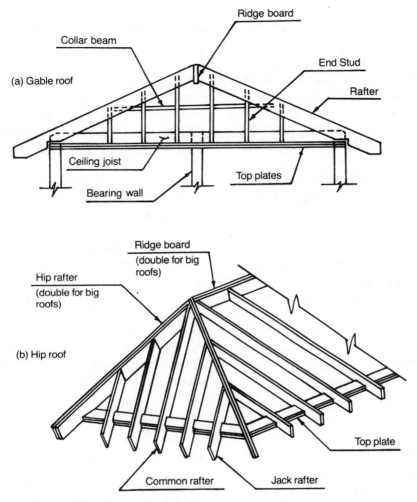

Figure 22.10 Gable and hip roof framing.

method is to *double* the ridge board and the hip rafters for gable and hip roofs.

Wall Sheathing

Wall sheathing is the exterior covering of the wall framing. It has several functions:

1. It provides a base on which the exterior siding is installed.

2. It provides the house with vertical brac-ing. This is very important in resisting wind and earthquakes.

3. It prevents air from filtering into the house.

4. It adds slightly to the thermal insulation of the walls.

So far, the most widely used material for wall sheathing is plywood panels. They are easy to install and strong in both directions, which makes them ideal for bracing. You may use sheathing grade, but engineering grade with exterior adhesive is recommended. The minimum thickness of plywood should be $\frac{3}{8}$-inch and $\frac{1}{2}$-inch for studs spaced at 16 and 24 inches o.c., respectively.

Plywood panels may be installed vertically or horizontally. In either method, the joints must be staggered in order to have continuous sheathing at any section. There should be a space of $\frac{1}{8}$ inch between the panels to allow for wood expansion. The edges of the panels must rest on and be nailed to framing members or two-inch thick blocking.

Particleboard panels are gaining popularity as sheathing material because they are inexpensive. Their thicknesses should be as recommended by the manufacturer. Other sheathing materials include **wood boards, insulating boards** and **gypsum boards,** but they are inferior to plywood or particleboard.

Roof Sheathing

Roof sheathing covers the roof framing. It serves the following purposes:

1. It forms a base on which the roofing material is installed.

2. It transfers the snow load to the rafters.

3. It braces the house against wind and earthquakes.

Plywood is the most popular material for roof sheathing. The quality of the material and the installation should be similar to those of wall sheathing. The direction of the face grain should be perpendicular to the direction of the rafters. The minimum thickness of plywood should be $\frac{1}{2}$- and $\frac{5}{8}$-inch for rafters spaced at 16 and 24 inches, respectively.

Board sheathing is also common. For asphalt shingle roofing, the boards should be laid close to one another. The boards may be **square-edged, shiplapped** or **tongue-and-groove.** Their minimum thickness should be $\frac{3}{4}$- or 1-inch for rafters spaced at 16 or 24 inches o.c., respectively. All joints should be made over the rafters. It is recommended that the joints of adjacent boards be staggered.

For wood shingle or shakes roofing, it is recommended to lay the boards with a space in between, to allow for ventilation. The boards are usually 1 × 3 or 1 × 4 inches. Their spacing (o.c.) should be equal to the exposure of the shingles. For example, for 16-inch shingles with 5 inches exposure, the boards should be 1 × 4 inches laid at 5 inches o.c.

Attic Ventilation

Attic ventilation removes the moisture that seeps into the attic from the house. It prevents condensation during cold weather and overheating during hot weather. The way attic ventilation systems work is that warm air moves upward, because it is lighter than cold air. Therefore, if an attic has two openings, one low and one high, the air will automatically move out of the attic through the high opening because the air in the attic is warmer than the outside air. This forces an equal amount of cool air into the attic through the low opening. This flow of air occurs even if there is no wind. The free area of vents should be about one square

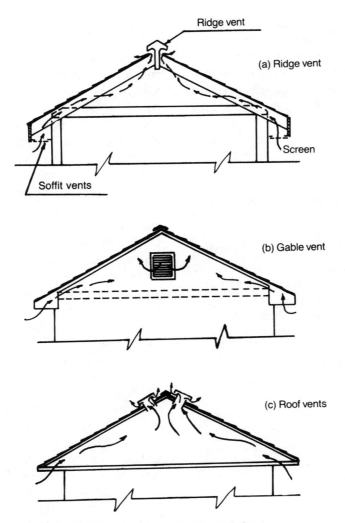

Ridge vent

(a) Ridge vent

Screen

Soffit vents

(b) Gable vent

(c) Roof vents

Figure 22.11 Different systems of attic ventilation.

foot per 300 square feet of attic. Attic ventilation systems may be:

Soffit and Ridge Vents In this system, the warm air moves out of the attic through a continuous ridge vent (Figure 22.11a). The replacement cold air is drawn in through equally spaced soffit vents. This system is good because it provides uniform flow of air throughout the attic.

Gable Vents In the gable vents system, the attic's warm air is discharged through vents located in the end gable (Figure 22.11b). The cold air is sucked into the attic through the soffit vents. This system, however, does not provide the attic with uniform ventilation.

Roof Vents Roof vents (Figure 22.11c) are used in hip roofs since ridge or gable vents are not applicable. The vents should be installed as high as possible in at least three sides of the hip roof. If an attic fan is provided, it replaces one of the vents. Roof vents are provided with overhanging covers to prevent rain water from getting into the attic.

Attic Fans

Installing an **attic fan** with a thermostat is highly recommended. On a typical sunny summer day, the temperature in an attic without a fan can exceed 150°F in the southern part of the United States and 120°F in the northern part. The thermostat is usually set at about 90°F. The fan operates automatically as soon as this temperature is reached. During hot or even warm sunny days, the fan works continuously for most of the day. This lowers the temperature of the house and significantly reduces the load on the air conditioning unit. Usually, you buy the fan, the carpenter installs it, and the electrician wires it. This type of fan costs a few hundred dollars.

Contracting for the Framing

In most cases, you buy the materials and contract labor only. Usually, the framing is contracted as a part of the overall carpentry, which includes the interior and exterior trimming, and installing the windows, doors, sky-lights, kitchen cabinets, vanities, wood siding and roofing. If several or all these items are contracted to one contractor, he or she should itemize the price of each item. This helps you in analyzing the bids and establishes the amounts of installments to be paid during the course of construction.

Before construction starts, you should shop around for lumber supplies. Mail a copy of the lumber list—as given in the bill of materials—to several suppliers, asking for their prices, the species and grade of the lumber and terms of payment. It is extremely important to deal with an honest and reputable supplier since lumber delivery is likely to take place during the day while you are not present to check the shipment. Many suppliers volunteer to make a **list of materials** from the drawings without charge as a service to their clients.

It is of interest to know that the labor cost of house framing is about equal to the price of the lumber. This should help you to estimate the contractor's price beforehand once you know the price of the lumber.

You can also estimate the cost of the carpentry by knowing the carpenter's productivity. For an ordinary frame, a carpenter can frame the walls and ceiling of a 75 square-foot area of the first floor or 60 square feet of the second floor in a day.

CHAPTER 23

Windows, Exterior Doors, and Exterior Trim

W indows and exterior doors should be installed as soon as the house frame is constructed in order to close the house and reduce theft and vandalism. Only windows and doors that can be installed during the day should be delivered because loose ones are an easy target for theft. Instruct the carpenter to take uninstalled windows or doors home with him for safe keeping until they are ready for installation.

Exterior trim includes window and door trim, porch, portico, posts, pediments, window shutters, and the cornice (eave overhang). A big portion of the exterior trim is sold in the form of finish lumber that is cut on site to fit various applications. Many items such as window shutters, posts, and pediments are shop-fabricated. These items are expensive and should be installed as soon as they are delivered.

Windows

The window styles and size (see Chapter 14) should be as shown in the design drawings. The size of each window is usually given in a specific designation. For example, a 3052 size double-hung window indicates that it has a nominal width of 3'-0" and a height of 5'-2". If the size is preceded by the letter M such as M 3052, it means that the window consists of two units installed side-by-side with a mullion in between. Visit several window showrooms to see the different styles so that you know the prices before the design drawings are finalized. Windows should be ordered in advance to guarantee their arrival on time for installation.

Windows consist of several components, and each has a designation. Figure 23.1—which is for a double-hung

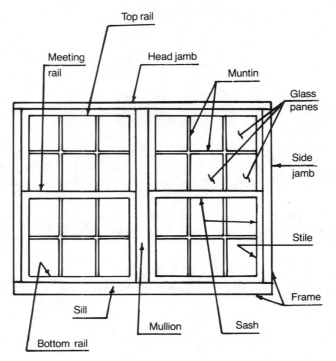

Figure 23.1 Window components.

window—gives an example of the most common window parts and their designations. They are:

Frame.   The exterior frame that holds the window together. It consists of two **side jambs,** a **head jamb** and a **sill.**

Mullion.   A vertical member that separates two side-by-side windows.

Sash.   The moving part of the window. It includes a frame consisting of two **side stiles,** either a **top or a bottom rail,** and a **meeting rail.** The frame holds the glass.

Muntin.   Short light bars that divide the glass into **panes.**

The sash of double-hung windows is balanced by springs that make it easy to operate and to hold it still in any vertical position. In most windows, panes are created by **separate vinyl grills** that are snapped into the window sash. This method is much cheaper and easier than muntins.

The window sash may be made of wood or aluminum. Wood is a better insulator than aluminum, but it requires painting or staining. Some manufacturers overcome this deficiency by incasing the exterior wood with vinyl. Wood windows may be made of Ponderosa pine, cedar, cypress, or redwood.

The windows must be installed perfectly horizontal and plumb. They should be securely nailed to the wood frame of the house, and should be covered with plastic or paper to protect them during construction. The exterior trim of the windows depends on the material of the exterior siding.

Exterior Doors

The exterior doors include the main entrance and side doors, and the doors to the back yard. The design drawings should include a list of the exterior doors, their sizes and types. Visit several showrooms before the drawings are finalized so that you may choose the doors that suit your family's taste.

The main components of a typical wood door (exterior and interior) are shown in Figure 23.2. For wear resistance, the door's sill is usually made of oak for wood doors and aluminum for metal doors. The sill is sloped outward in order to prevent the rain from entering the house. Most metal doors are sold already weatherstripped.

Main Entrance and Side Doors The main entrance door should be given some thought since it has a dramatic impact on the appearance of the house, and thus affects its resale value. Typical doors are $1\frac{3}{4}''$ thick, 6'-8" high, and 3'-0" wide. The main entrance door of a big house is likely to consist of a pair of three-foot wide doors. The door may be made of steel or wood. Steel doors are fireproof, light, maintenance free, and have a thermal insulating core. Wood doors, on the other hand, are heavy and have less thermal insulation characteristics, but they can be beautiful, particularly when professionally stained. Wood doors may be made of Ponderosa pine, cedar, cypress, or redwood.

The basic exterior trim for main entrance doors is the casing. It consists of decorative strips that cover the gaps between the door frame and the surrounding house frame. More elaborate trim may include a portico, pilasters, or a pediment.

Side doors are mostly steel, with or without panel lights. They have the same thickness and height as the main entrance door but their width is 2'-8".

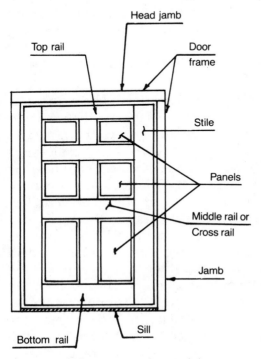

Figure 23.2 Components of a panel door.

Head jamb

Top rail

Door frame

Stile

Panels

Middle rail or Cross rail

Jamb

Sill

Bottom rail

Back Yard Doors Most doors leading to the back yard are either of the sliding or French variety. The sliding panel is provided with rollers for smooth operation. Quality doors have vinyl sheathing encasing the exterior wood for better appearance and reduced maintenance.

Exterior doors, particularly those made of wood, should be protected during construction so that they do not get dented or stained. For security, only one door should be used for access. The rest should be blocked and covered with plywood boards or heavy

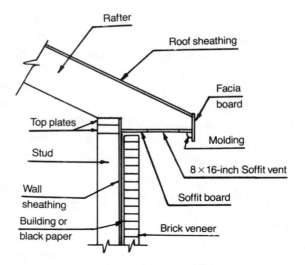

Figure 23.3 Narrowbox cornice.

plastic sheets. As soon as they are installed, wood doors should be given at least one coat of clear wood preservative to protect them from moisture.

The Cornice

The cornice is the covering of the eave overhangs. The most common cornice types are the narrow box and the wide box.

Narrow Box Cornice   The narrow box cornice extends beyond the outer face of the exterior walls by up to 16 inches. The details of a narrow box cornice for a house with brick veneer exterior siding are shown in Figure 23.3. The major components of the narrow box cornice are the **facia** and **soffit boards.** They may be made of plywood, hardboard, or aluminum. Soffit boards are cut to accommodate soffit vents consisting of openings covered with metal shutters. The size and spacing of the openings should be shown in the design drawings.

It is recommended to paint wood facia and soffit boards with at least one primer coat before they are installed. This way, it is easier and less costly. In your planning, take into consideration that wood trim requires painting every four years at a considerable cost, while aluminum trim is maintenance free.

Wide Box Cornice   The wide box cornice is more elaborate than the narrow box. In addition to the facia and soffit boards, wide box cornice has **lookouts, frieze boards** and **headers** (Figure 23.4). The lookouts are horizontal joists with the outer ends fastened to the rafters and the inner ends to the frieze boards. Their function is to support the soffit boards. Headers are fastened along the outer ends of the lookouts and rafters. They enclose the cornice and provide support for the facia boards. The materials, maintenance, and soffit vents should be the same as for the narrow box cornice.

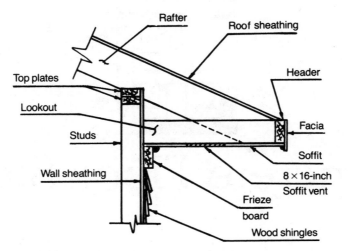

Figure 23.4 Wide box cornice.

Contracting for Windows, Doors, and Exterior Trim

It is cheaper to buy the windows, doors, and the materials of the exterior trim from material suppliers, and contract installation only. It is important to buy high quality windows and doors since they look better and last longer.

Virtually all windows and glass doors have double insulating glass as well as screens. Some of them carry a five-year warranty. The installation of the windows, doors, and exterior trim is usually contracted to the carpenter who constructs the house frame.

CHAPTER 24

Concrete Floor Slabs

Concrete floor slabs have several applications in house construction: (1) basement, (2) garage, and (3) above grade living areas. The structural integrity of concrete floor slabs depends on the strength of the concrete slab, **the base** on which the slab is poured, and the **subbase** on which the base is laid. If any of these elements is weak, the slab will crack and deform irregularly.

Subbase

The subbase is the soil on which the base is laid. It should be either undisturbed soil or clean fill. In either case, the soil should be well drained. When fill is used, it should be laid in six-inch layers. Each layer must be well compacted (an efficient method of compacting fill is by saturating each layer with water). The soil should be as uniform as possible so that it settles evenly. The subbase should not include organic materials (for types of soils, see Chapter 19).

Base

The base consists of a four- to six-inch thick layer of gravel or crushed stones. It serves two functions: (1) it distributes heavy concentrated slab loads on a wide area, and thus reduces the chance that the slab will crack, and (2) it breaks the **capillary attraction,** a natural phenomenon whereby subsurface water is drawn upward in the soil by a distance proportional to the fineness of the soil particles. In clay, this distance can be as high as 10 feet; in gravel it is zero.

Vapor Barrier On top of the gravel, a vapor barrier, also called moisture barrier, consisting of four- to six-mil (1.0 mil = 1/1000 inch) thick polyethylene plastic sheets should be laid. Without the vapor barrier, the water vapor in the soil will penetrate and condense on top of the concrete slab, which increases the level of humidity. If the floor is covered with resilient tiles, the water vapor will condense under the tiles causing them to loosen.

Insulation Concrete slabs lose heat by direct contact with the soil on which they are poured and along their edges. Basement slabs are poured a few feet below grade. At this elevation, the soil's temperature tends to be relatively constant all year round (55°F on average). The temperature differential between 55°F and 70°F does not justify insulating the basement slab.

For floor slabs poured above the grade elevation, the soil close to the edges of the house gets cold and this causes the slab to lose heat. Most of the heat loss occurs within a three-foot wide strip around the perimeter of the slab. This is in addition to the heat loss through the edges of the slab. An efficient layout for above-grade slab insulation is shown in Figure 21.9b.

The insulation should be of the rigid type. It must be strong enough to withstand the weight of the slab and the dead and live loads above it. Its R-value must meet the state and local code requirements.

The Concrete Slab

The concrete slab is poured on top of the vapor barrier or insulation. It should not be less than four inches thick. 3000 PSI concrete is recommended. In order to control cracking and to resist uneven settlement of the subbase, the slab should be reinforced by a 6 × 6 – #10 × #10 wire mesh placed at the middle of the slab's height.

For uninsulated basement slabs, the joint between the edges of the slab and the foundation wall should be filled with asphalt to prevent subsurface water from leaking into the basement. The garage floor slab should be pitched by about two inches from the back to the front of the garage, in order to prevent any water from accumulating in the garage. An on-grade slab may be heated by an embedded heating duct (see Figure 21.6c).

Contracting for Concrete Floor Slabs

The unit by which a concrete floor slab is contracted is the square foot. This includes subbase preparation, gravel or crushed stones, moisture (vapor) barrier, insulation if any, wire mesh, and pouring and finishing the concrete. The cost of labor can be as much as 150 percent of the cost of materials. This is an area where you can save some money if you have the help of unskilled laborers, who will do most of the work, and a mason or two to finish the surface of the concrete. One mason can finish 1000 square feet of concrete slab per day.

CHAPTER 25

Roof Covering, Gutters and Downspouts

Roof covering provides the roof with a waterproof finish in order to protect the house from rain, snow, and wind. The major components of the roof covering are: (1) **underlayment,** (2) **surface materials,** and (3) **flashing.**

Gutters and downspouts make up the drainage system of the roof. They protect the eaves of the roof, exterior walls, and basement from water damage caused by rain or melting snow.

Underlayment

The underlayment is a moisture-resistant layer made of **felt saturated with asphalt.** It is applied over the roof sheathing or decking to provide a smooth dry surface for laying the surface materials.

Underlayment is manufactured in 15 and 30 pounds per **roof square** (1.0 roof square = 100 square feet of roof surface) and is sold in 36-inch wide rolls. The 15-pound underlayment is used under asphalt shingles, and the 30-pound underlayment is used under tile and slate and between the courses of wood shakes. Wood shingles do not require underlayment except at the eaves in heavy snow climate.

Before installing the underlayment, a **drip edge** is fastened along the eaves (the bottom edges of the roof). It consists of a metal strip—usually aluminum—with a short flange. The drip edge is laid such that the flange faces out and down. The underlayment is then laid in horizontal layers starting from the eaves and going up. Each layer overlaps the preceding one by two inches. The overlap is four inches in the horizontal direction and six inches over the ridges and hips. Next, a drip edge is installed on top of the underlayment along the **rakes,** which are the sloped ends of the gable roofs. Underlayment should not be left uncovered for more than two hours, since the sun causes it to wrinkle.

In heavy snow areas, it is a good idea to install a 36-inch wide layer of 45-pound roll roofing along the eaves above the drip edge and below the underlayment. Its purpose is to protect the roof from the effects of the **ice dams** that form in the gutters and along the eaves when the weather warms up and then freezes again after a heavy snowfall. Ice dams block the free discharge of water in the gutters causing it to back up under the shingles.

Surface Materials

The surface materials form the finish cover of the roof. The most popular surface materials are asphalt shingles, asphalt roll roofing, wood shingles, wood shakes, tiles and slate.

Asphalt Shingles Asphalt shingles are very popular because they are inexpensive and easy to install. They consist of a **felt** or **fiberglass** base saturated

with asphalt on both sides. In addition, the exposed surface is coated with colored **mineral granules** for protection and appearance. The fiberglass-base shingles cost a little more than felt-base but they are fire resistant and more durable. Fire resistant felt-base shingles are available, but they cost more than comparable fiberglass-base shingles. Both the felt and fiberglass shingles are sold in assorted colors, though the former comes in a greater variety. The color you choose for the roof shingles should be compatible with the color of the wall siding.

Most asphalt shingles are sold in the form of 12 × 36-inch strips called **three-tab** or **square-tab** shingles (Figure 25.1a). Of the 12-inch side of the strip, only five inches is exposed. The rest is overlapped by the following course. Each shingle strip has a **self-sealing**

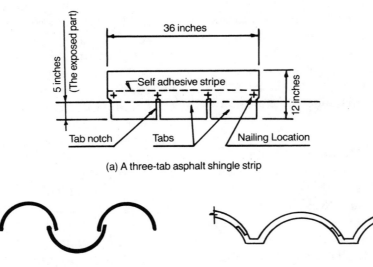

(a) A three-tab asphalt shingle strip

(b) Mission tiles   (c) Spanish or S tiles

(d) Greek tiles

Figure 25.1 Roof surface materials.

stripe that makes the shingles interlock and settle down under the warmth of the sun. This eliminates wrinkling and increases the shingles' resistance to strong winds, rain, and snow.

Asphalt shingles are classified according to their weight per **square** (1.0 square of shingles = 100 square feet of exposed surface). The most popular weights are 235, 300 and 325 pounds. Shingles are sold in bundles, each containing 27 strips, three bundles to a square.

Before starting installation, the roof is measured in order to lay out the shingle courses and the pattern of the tabs. The edge of each course is marked on the underlayment by snapping a horizontal chalkline. The position of the first strip of each course is marked by a perpendicular chalkline.

Installation begins with laying a five-inch wide **starter-strip** along the eave edges. Its function is to underlay the notches in the first course shingles. The courses are laid starting from the eaves going up.

The roof ridges and hips are capped by special curved shingles. These shingles should be installed after all plane surfaces are completed.

Asphalt Roll Roofing   Asphalt roll roofing is made of felt covered with heavy layers of asphalt. It is sold in 36-inch wide rolls. Roll roofing generally weighs between 45 and 90 pounds per roof square. It is manufactured in two forms: **smooth surface** and **mineral surface.** The first type is less expensive and easier to install. The second type is made by coating the surface of the material with colored mineral granules.

Roll roofing is a one-layer installation, meaning it functions as both underlayment and surface materials. This makes it economical. It is mostly used in rural areas. Its biggest disadvantage is its short life span; some types are guaranteed for five years only.

Roll roofing may be installed in horizontal or vertical layers. The method of installation should be as recommended by the manufacturer.

Wood Shingles   Most wood shingles are made of western red cedar or redwood. They are all-heartwood, all-edge grain,* and machine sawn tapered. Wood shingles are cut in 16-, 18- and 24-inch lengths and random widths. In houses with normal pitched roofs, the lengths used most are 16 and 18 inches of which 5 and 6 inches are exposed. Roof shingles should be **No. 1 grade.**

Wood shingles are more expensive than asphalt shingles. They also do not resist fire unless chemically treated. Some local building codes require that wood shingles be fire rated. The price of fire rated shingles is about 75 percent higher than unrated ones.

For the most part, wood shingles are installed on open board decks. However, some building codes require that shingles be laid on solid sheathing. In this case, the roof sheathing should be covered with a layer of underlayment. The shingles are installed over a series of **furring strips** laid horizontally at a spacing equal to the shingle exposure.

Wood shingles should be laid staggered with $\frac{1}{8}$-inch spacing between adjacent shingles to allow for expansion. Each shingle should be fastened by only two nails to reduce its chances of splitting.

---

*U.S. Department Of Agriculture, Agriculture Handbook No. 73.

Wood Shakes   Wood shakes are made of the same wood as wood shingles, but they are hand split. In comparison with wood shingles, wood shakes are longer and much thicker, their surface is rougher, and their price is higher. The lengths used most are 18 and 24 inches of which 7½ and 10 inches are exposed.

Wood shakes installation is similar to that of wood shingles. One main difference is that a layer of 30-pound underlayment is installed in between successive courses.

Clay Tiles   Clay tiles are identified with several house styles: Mission, Mediterranean, Italian Villa, and Spanish Villa. They are popular in California, Florida, and Arizona. There are three main types of clay tiles: Mission, Spanish, and Greek.

Mission tiles are the most popular. They are red and are semicircular in shape. The tiles are laid normal to the roof slope in alternate rows. Rows with the tile openings facing up are overlapped by the adjacent rows in which tile openings are facing down (Figure 25.1b).

Spanish tiles are S shaped. They are also laid in rows normal to the roof slope. The tile edges of each row overlap the tile edges of the preceding row (Figure 25.1c).

Greek tiles are perhaps the most expensive (their price is about 15 times as much as the 235-pound asphalt shingles). The tile pattern consists of pans laid in rows normal to the roof slope with a space in between. This space is bridged by cover tiles laid face down (Figure 25.1d).

Clay tiles are fireproof and very durable. But they are heavy, expensive, and difficult to repair if cracked. They must be installed by specialized professionals. The roof framing and sheathing must be designed to carry the heavy weight of the tiles.

Concrete Tiles   Concrete tiles are gaining popularity over clay tiles because they are lighter, cheaper, and easier to install. They are manufactured in curved or flat shapes. The roof frame and decking should be designed to support the heavy weight of the tiles.

Slate   Slate is a natural stone, sliced into shingles of random sizes. Like clay tiles, slate is heavy, beautiful, durable, fire resistant, difficult to repair if cracked, and very expensive. Slate is available in gray, green, black, red, and other colors.

Flashing

Flashing is required in places where regular roof covering does not provide adequate waterproofing: (1) the valley where two sloping roof planes meet, (2) around the chimney, (3) where the roof meets a vertical wall, (4) around skylights, and (5) around vent pipes. Valley and chimney flashing are the most common.

Valley Flashing   Valley flashing may be made of metal (tin or galvanized steel) or asphalt roll roofing.

Metal flashing used in normal pitch roofs is usually 18 inches wide. It has a channel or a recess along its centerline to facilitate speedy discharge of water. The flashing edges are bent upward to prevent water from flowing over the adjacent shingles.

Installation starts with laying 36-inch

wide asphalt roll roofing centered on the valley. This layer is subsequently covered by the roof underlayment. Then, the metal flashing is installed on top of the underlayment. Finally, the roof shingles are laid such that they cover a part of the flashing, leaving the other part exposed. At the valley's top, the width of the exposed part is equal to three inches on each side of the central channel. This width increases by $\frac{1}{8}$-inch per foot going down along the valley (Figure 25.2).

Roll roofing flashing may be open valley or closed valley. In the open valley method, a 36-inch wide strip of 90-pound roll roofing is used as flashing. The roof shingles cover part of the roll roofing leaving the other part exposed (similar to metal flashing). In the closed valley method, the asphalt shingles of the two intersecting planes of the roof overlap, and cover the roll roofing flashing completely. This method is recommended because it is more durable.

Chimney Flashing   Chimney flashing covers the joint between the roof and the chimney. It consists of two parts: (1) an angle-shaped metal **base flashing** with one leg laid underneath the roof shingles and the other leg laid along the chimney side, and (2) a metal **counterflashing** the function of which is to cover the base flashing and the joint. It has one side laid over the base flashing and the other side inserted in the chimney's mortar joints (Figure 25.3a). The joints are caulked to make them waterproof.

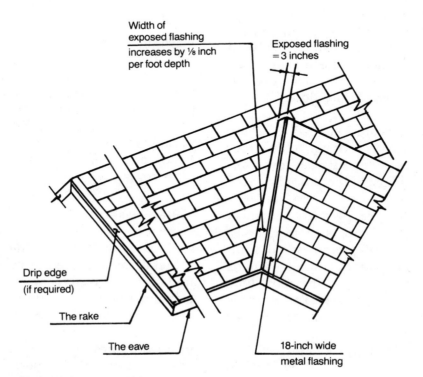

Figure 25.2 Metal roof flashing.

For wide chimneys, a wood saddle is constructed along the high side of the chimney over the underlayment (Figure 25.3b). Its purpose is to prevent water from accumulating behind the chimney. The saddle is covered with base flashing, then by roof shingles, followed by counterflashing.

Gutters and Downspouts

The gutters, also called troughs, collect the water along the eaves and discharge it into the vertical downspouts. In turn, the downspouts, also called leaders, discharge the water into a drainpipe or on concrete splash blocks. The layout of gutters and downspouts should be shown in the design drawings.

Materials Both gutters and downspouts may be made of aluminum, vinyl, steel, copper, or wood:

Aluminum. Aluminum is very popular because it is corrosion resistant, inexpensive and easy to install. Gutters and downspouts are sold in standard parts to be assembled on the job. Aluminum gutters and leaders with enamel finish are available in two colors: white and brown. Gutters are sold in two thicknesses: 0.027 and 0.032 inches. Downspouts are sold in two sizes: 2 × 3 inches and 3 × 4 inches; and two thicknesses: 0.020 and 0.025 inches. Generally, the light-gage aluminum is guaranteed for 15 years, and the heavy gage for 20 years.

Vinyl. Vinyl gutters and downspouts are gaining in popularity because they are easy to install, and do not rust or chip. They are sold in a limited variety of colors; white is the most popular and least expensive.

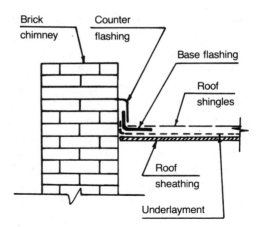

(a) Base and counter flashing

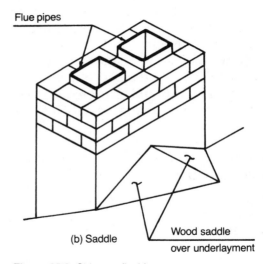

(b) Saddle

Figure 25.3 Chimney flashing.

Steel. Steel gutters and downspouts with enamel or galvanized finishes are available. They have a short life span and are prone to rust, requiring periodic painting and maintenance. This makes them unpopular, even though they are inexpensive.

Copper. Copper is very durable, but very expensive.

Wood. Wood gutters are beautiful, but they require frequent treatment with water-repellent preservative. This makes them unpopular.

Installation   Gutters should be pitched toward the downspouts to accelerate the water flow. Metal gutters may be fastened to the roof sheathing or decking by strap hangers or to the facia boards by spikes. Fasteners should be spaced at 30 inches o.c., maximum. Aluminum gutters should be fastened only by aluminum nails or screws to avoid chemical reaction with other metal. The joints should be caulked to prevent leakage.

Wood gutters are fastened to the facia boards by rust proof screws. Furring blocks are placed between the gutters and facia boards in order to prevent moisture from accumulating in between them.

Downspouts are fastened to the exterior siding and the backing studs. A strip fastener should be provided every five to six feet of the straight part of the downspout.

Contracting for Roofing, Gutters, and Downspouts

Contracting can be split into two parts: (1) roofing and (2) gutters and downspouts.

Roofing   Roofing may be contracted either material and labor or labor only. Either a roofer or the framing carpenter can install the asphalt or wood shingles. Contracting may be per square or on a fixed price basis for the entire roof.

On average, the labor cost of installing asphalt shingles is equal to the price of the shingles. Another method of estimating the labor cost of roof covering is by calculating the hours required to complete the job. A roofer or carpenter can install about four squares on a one-story high roof or three squares on a two-story high roof per day.

For wood shingles and shakes, the labor cost is about equal to the price of the shingles. A roofer or carpenter can install 2.5 squares on a one-story high roof or 2 squares on a two-story high roof per day.

Gutters and Downspouts   Gutters and downspouts are usually assigned material and labor to a roofer. The labor cost is almost the same for heavy- or light-gage aluminum. Two roofers can install about 180 linear feet of gutters or 250 linear feet of downspouts per day.

CHAPTER 26

Garages and Garage Doors

Garages may be **attached** (to the house), **detached** (separate), or **carport,** which is a shed with a roof but without sidewalls. Carports may be either attached or detached.

The size of the garage has become an important factor in determining the price of a house. As a result, more houses are being built with two- or three-car garages than ever before.

The location of the garage door is getting more attention from architects, building designers, and homeowners. Garage doors are often located on the side of the house, while the front of the garage is provided with windows that match those of the house. This design makes the garage look like a living area, adding prestige to the house.

The Size of the Garage

The depth of the garage should not be less than 20 feet. Its width should not be less than 13, 20, and 27 feet for one-, two-, or three-car garages, re-

spectively. Workshop and storage areas are to be added to these dimensions.

Garage Door

Garage doors are either single or double. The single is about 9 feet wide and 7 feet high. The double is about 16 feet wide and 7 feet high. A two-car garage may have either a double door or two single doors. A three-car garage may have either three single doors or one single and one double door.

Most garage doors are of the sectional type, which consists of four or five horizontal sections attached to each other by hinges. This allows the doors to bend along the curved sidetracks when they open or close. Some doors are paneled (Figure 26.1); others have glass panes in the top section (Figure 26.2). Many homeowners like the garage door to match the surrounding exterior siding in order to make the front uniform (Figure 26.3).

Garage doors may be opened man-

Figure 26.1 Paneled garage door.

Figure 26.2 Garage doors with glass panes.

Figure 26.3 A matching garage door.

ually or automatically by remote control. The manual doors are usually of the single type since double doors are heavy to lift.

The construction of the foundations and wood frame of the garage is similar to those of the house. The framing of carports is usually post and beam. The header of a single-garage door is likely to be made of lumber. The header of double-garage door is likely to be made of a steel beam that is supported at each end by steel pipes.

Contracting for a Garage Door

The garage door(s) should be contracted before construction starts so that the fabricator is ready to take the opening measurements and construct the door as soon as the framing and concrete slab are constructed. It is very important to have the garage door installed by the time the house frame is completed in order to close the house.

Door fabricators have catalogues showing a wide variety of designs and styles. In order to make the right choice, invite several door fabricators—one at a time—to the site and take them on a tour of the area to check on different styles. For doors that match the exterior siding of the wall (Figure 26.3), the door fabricator provides the frame of the door and you provide the siding lumber. The exterior siding of the surrounding wall should be installed after the garage is completed in order to match its pattern.

For remote control operated doors, the automatic operator is installed after the electricity is connected. In the interim period, the door can be locked from the inside by a latch that operates manually.

CHAPTER 27

Exterior Siding

*E*xterior siding is the covering of the exterior walls. Its function is to protect the house from the weather and to give it an attractive appearance. It is available in wood, metal, vinyl, masonry (brick and stone veneer), and stucco. In choosing the material of the exterior siding, consideration should be given to initial cost, maintenance, fire resistance, appearance, and your family's taste. Installation of exterior siding includes covering the wall sheathing with a layer of sheathing paper (p. 234).

Wood Siding

The wood used for exterior siding should have good resistance to weather-caused defects such as warping, splitting, and decaying. It should also be receptive to painting and staining.

High quality* exterior siding woods include cedar, eastern white pine, sugar pine, western white pine, cypress, and redwood. Good quality woods include western hemlock, ponderosa pine, spruce, and yellow poplar. Fair quality

woods include Douglas fir, western larch, and southern pine.

Wood siding may be installed in horizontal, vertical or diagonal patterns.

Horizontal Patterns Horizontal patterns include:

Bevel Siding. Bevel siding, also called **clapboard,** consists of long boards of tapered cross section (Figure 27.1a). Board widths vary from 4″ to 10″, and their butt thicknesses from $\frac{1}{2}$″ to $\frac{3}{4}$″. The thickness of the thin edge of a board is $\frac{3}{16}$″, regardless of its size. The boards are laid starting from the bottom course. Each course should lap the preceding one by at least one inch. The boards should be nailed to the studs through the wall sheathing using corrosion resistant nails. Board joints should be staggered.

Drop Siding. In drop siding, also called rustic siding, the thickness of the boards is uniform. The most popular board sizes are 1″ × 6″ and 1″ × 8″. The edges of the boards may be **tongue-and-groove** or **shiplap** (Figure 27.1b). The former can resist rain penetration better than the latter. Drop siding may be used for garages and secondary buildings with-

---

*U.S. Department of Agriculture, Agriculture Handbook No. 73

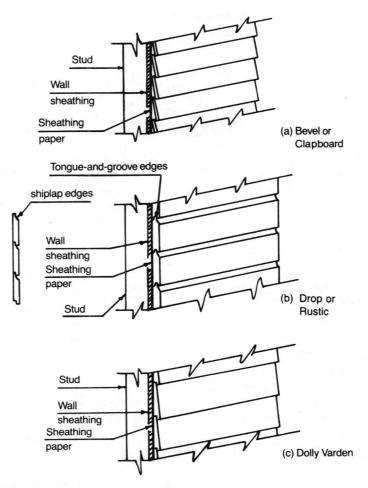

Figure 27.1 Horizontal pattern siding.

out sheathing. However, the frame of the building must be provided with diagonal bracing in order to resist the horizontal forces of wind and earthquakes.

Dolly Varden Siding. The boards of Dolly Varden siding are tapered and their edges are shiplapped (Figure 27.1c). This siding may be used for garages and other secondary buildings without sheathing, provided that the wood frame is diagonally braced.

Plywood Lap Siding. Plywood boards of sizes $\frac{3}{8}'' \times 12''$ or $\frac{3}{8}'' \times 16''$ are laid lapped. The lap should not be less than one inch. Board joints should be backed by shingle wedges to which the adjoining boards are nailed.

Vertical and Diagonal Patterns The vertical and diagonal patterns of board siding are **board and batten, batten and board,** and **board and board** (Figure 27.2). The sequence of their identification follows the sequence of their installation. Drop siding may also be used in vertical and diagonal pat-

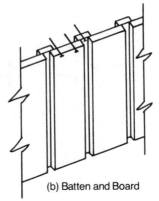

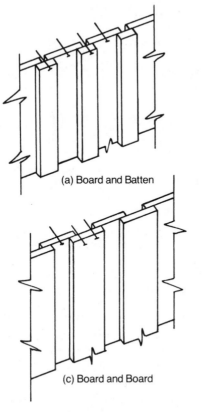

Figure 27.2 Vertical and diagonal pattern siding.

terns. The thickness of plywood sheathing should not be less than $\frac{5}{8}''$, and that of board sheathing should not be less than one inch.

Plywood panels are popular because they have a wide range of finishes. They are sold in 4 × 8-, 4 × 9-, and 4 × 10-foot sizes, which makes them easy to install. Plywood panels are available in different grades and thicknesses.

Wood Shingles and Shakes

Wood shingles and shakes are popular because they are relatively cheap, durable and easy to install. Shingles and shakes made of the heartwood of west-

ern red cedar, northern white cedar, bald cypress, or redwood are rot and insect resistant and do not require painting or staining unless you want to change their color. They will weather in a silver grayish color.

The shape and size of the shingles are similar to those of the roofing except that they are of lower quality. Siding shingles may be grade No. 1, 2, 3, or undercourse grade, depending on the method of installation.

Installation There are two methods of installing wood shingles and shakes: (1) **single-course** and (2) **double-course.** In both methods, the shingles and shakes may be installed over either

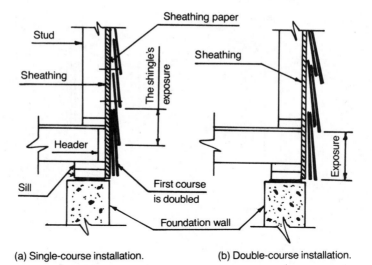

(a) Single-course installation.   (b) Double-course installation.

Figure 27.3 Shingles and shakes siding.

board or plywood sheathing. In the single-course method, (Figure 27.3a), each course is overlapped by the following one, similar to roof application. The maximum exposure is $7\frac{1}{2}''$ and $8\frac{1}{2}''$ for 16″ and 18″ long shingles, respectively. Adjacent shingles should have a space of $\frac{3}{16}''$ in between to allow for expansion. The shingles used in this method are usually grade No. 2.

In the double-course method, two courses are laid on top of one another (Figure 27.3b). The exposure is 12″ and 14″ for 16″- and 18″-long shingles, respectively. The bottom course may be grade No. 3 or undercourse grade. The top course should be grade No. 1. In either method, it is recommended to apply a coat of wood preservative to the shingles or shakes before installation.

Metal Siding

Metal siding may be aluminum or steel (the former is more popular). It is sold in the form of extruded strips with flanges so that adjacent strips can be locked together. The strips are nailed to the sheathing through elongated holes punched in the top of the strips to allow for the free expansion and contraction of the siding. Metal siding is popular because it is fire resistant, lightweight and immune to attack by termites. Aluminum siding is available in baked enamel or plastic finishes, and therefore does not require painting. It is sold in different colors, white being the most popular. Most aluminum and steel siding is guaranteed for about 35 years.

There are some disadvantages to metal siding:

1. It is a good electrical conductor. It must be grounded by connecting the siding to a water pipe with a cable.

2. It is a good heat conductor. In summer, it absorbs heat from the sun and radiates part of it into the house. In winter, it has little resistance to heat loss (some types are backed by a thin layer of insulation).

3. It is noisy in the rain.

4. It dents easily. This is particularly true of aluminum.

Vinyl Siding

Vinyl siding is sold in the form of strips, similar to those of metal siding. It does not rust, and therefore does not require painting. Vinyl resists denting better than aluminum. However, it tends to be brittle at low temperatures, and may twist under strong sun. Vinyl expands and contracts markedly with temperature changes. Therefore, it is nailed to the sheathing through elongated holes punched in the top edge of the siding.

The exterior trim that includes window and door trim and the cornice may be made of special molded vinyl strips. More popular however, is to use aluminum strips that are formed on site to fit each application. The ends of the siding should slide under the trim without being fastened to it in order to allow the siding to expand and contract freely.

Brick and Stone Veneer

Brick and stone veneer siding gives the house the appearance of being built of solid bricks or stones. The fact is that veneer is merely a decorative cover installed over the wall sheathing. It does not support any load other than its own weight. Brick or stone veneer has several advantages: (1) it is beautiful, (2) it is maintenance free, (3) it increases the resale value of the house, and (4) it has better insulating characteristics than other types of siding.

Bricks and Stones Bricks are made in different sizes, colors, and textures. The most common size is the standard, which is $8 \times 4 \times 2\frac{1}{4}$ inches (nominal dimensions). Other sizes include Norman, Roman, Jumbo, and Norwegian. There is also hand-made brick, which has irregular surfaces. Stone used in veneering consists of 4-inch thick slices of limestone or sandstone. They may be cut in rectangular or random shapes. Both brick and stone veneer are expensive. In many houses, they are used to cover only the front of the house or a part of it. The rest of the house is covered with less expensive siding. Visit several brick suppliers early on in order to check on the variety and the prices. You may also wish to consider used bricks.

Flashing and Sheathing Paper A **base flashing** consisting of either corrosion-resistant metal or 30-pound asphalt felt should be installed at the brick course below the bottom of the sheathing (Figure 27.4). Weep holes should be provided every four feet of the same course to discharge any moisture that may accumulate behind the bricks.

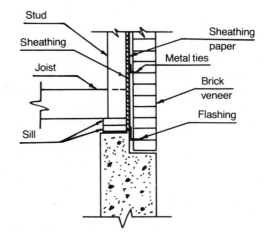

Figure 27.4 Brick veneer construction.

Sheathing paper, also called **black paper, building paper,** or **breathing paper,** consisting of 15-pound asphalt felt rolls is installed on the sheathing starting from the top of the flashing going up. It prevents the outside moisture from passing through to the sheathing and wood framing but allows the water vapor of the house to pass through to the outside. Sheathing paper should be installed horizontally. Each course should lap over the preceding one by four inches, minimum.

Brick and Stone Veneer Construction   Brick and stone veneer bears on either a shelf in the foundation wall or a separate concrete block wall (see Chapter 21). The bricks or stones are capable of supporting themselves in the vertical direction, but they are weak in the lateral or horizontal direction. Lateral support is provided by tying the walls to the wood sheathing by corrosion-resistant metal ties spaced every three feet in the horizontal direction and about every 16 inches in the vertical direction.

Brick and stone veneer walls do not expand or contract if the ambient humidity increases or decreases, but the wood framing to which they are tied does. The relative expansion or contraction exerts a considerable amount of pressure on the brick or stone walls and may cause them to crack. In this regard, balloon framing is superior to platform framing (see Chapter 22) in that its shrinkage or swelling in the vertical direction is negligible.

Stucco

Stucco is made of cement, sand and water. It may or may not include lime. Stucco siding is a characteristic of the Elizabethan Half-Timber and Spanish house styles (see Chapter 14).

Stucco siding starts with installing a layer of expanded metal lath over the sheathing paper. Its function is to prevent the stucco from cracking when the house frame expands or contracts relative to the stucco. The stucco is then applied in three coats: (1) **scratch coat,** (2) **brown coat,** and (3) **finish coat.**

Scratch Coat. The scratch coat is about $\frac{3}{8}$" thick. The stucco should be pressed behind the metal lath so that it covers the lath completely. While the stucco is still in the plastic state, it is scratched horizontally by a rake-shaped tool in order to create a good bonding surface. The scratches must not expose the metal lath.

Brown Coat. The brown coat is also about $\frac{3}{8}$" thick. This coat must produce a perfectly plane surface. Any imperfection in this coat will show after the finish coat is applied. The surface should be roughened in order to bond to the finish coat.

Finish Coat. The finish coat should not be less than $\frac{1}{8}$" thick. It can be made in many decorative styles. The texture and color of this coat is up to your taste. The mason should show you a brochure including different finishes to choose from.

Contracting for Exterior Siding

Exterior siding may be contracted either labor and material or labor only. It is interesting to note the price of the materials even if you intend to contract the job labor and material, because this helps you estimate the labor costs. For wood siding, the labor cost should be slightly less than the price of the material. The labor cost of installing wood shakes and shingles is about equal to the cost of the materials. The labor cost of installing uninsulated aluminum siding is about 20 percent higher than the cost of the materials. Your contract with the brick veneer contractor should include a clause that the contractor must clean the brick off before the mortar hardens. It should be understood that these guidelines for prices are subject to local conditions in the construction market.

CHAPTER 28

Central Heating and Air Conditioning

Central heating systems may be classified as **forced hot water** or **forced hot air** according to the medium that is used to transfer the heat from the heating plant to different parts of the house. Forced hot water circulates through small water pipes, while forced hot air circulates through big air ducts. **Central air conditioning** systems are classified as **forced air** since they all use air to transfer coolness from the cooling units to different parts of the house.

If you are installing central air conditioning, you ought to consider hot air heating since one duct system can serve both.

The heating plant for hot water heating systems may be an oil or gas boiler. The heating plant for hot air heating may be an oil, gas, or electrical furnace. All central air conditioning units run on electricity.

Heat pumps are electrical units that deliver heating or cooling by the flip of a switch. Their medium of transference is forced air.

Solar heating systems are getting some attention, but they haven't made much headway because of their high initial cost. It is more economical to utilize the sun's energy for heating water.

Hot Water Heating System

A hot water heating system consists of a **boiler,** which may be fired by oil or gas, an **oil** or **gas feeding pipe**, a **water feeding pipe**, a **pressure reducing valve**, a **supply** and a **return** water pipe for each heating zone, a **circulator pump** for each heating zone, an **expansion tank** for the entire system, and **baseboard heaters** and/or **radiators** for the areas to be heated (Figure 28.1). A hot water heating system can accommodate up to **six** heating zones.

Heating of a particular zone starts with a signal from the thermostat that switches on the circulator pump and the boiler. The pump forces the water of the return pipe into the boiler where it gets heated. It leaves the boiler in the supply pipe to the baseboards and/or radiators where it emits a part of its heat. The water then returns to the boiler through the return pipe (Figure 28.1), where it gets heated once more. This cycle is repeated until the temperature of the area being heated reaches the preset temperature of the thermostat. The thermostat then sends another signal that switches off the circulator pump and the boiler.

Usually, the water leaves the boiler at about 200°F and returns at about 180°F. For more comfort, the system may be designed to deliver the water at 180°F and return at 160°F. This can be accomplished by increasing the length of the baseboards and the size of the radiators.

Boiler The boiler is the heating plant of hot water systems. It may be fueled by gas or oil, depending upon the availability of gas in the area. Gas boilers do not accumulate deposits and thus do not require frequent maintenance. Oil boilers, on the other hand, require an annual tune-up, otherwise their efficiency drops.

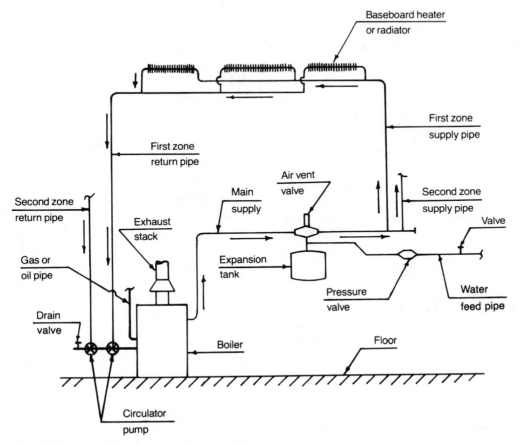

Figure 28.1 A two-zone hot water heating system.

Water Feed Pipe The water feed pipe provides the system with the water needed for its operation. It has a **feed water valve** that can shut off the water flow to the system in case of repair. The water feed pipe is connected with a **pressure reducing valve** which automatically relieves the water pressure if it exceeds a certain level.

Supply and Return Pipes A supply pipe carries the heated water from the boiler to the **baseboards** and **radiators** of each heating zone. A return pipe carries this water back to the boiler to be heated once more.

Circulator Pump Each heating zone has a circulator pump to force the hot water of the zone into circulation. The pump is switched on and off by the thermostat.

Expansion Tank The expansion tank has three functions: (1) it absorbs the expansion and contraction of the water due to temperature changes, (2) it regulates the pressure of the water to prevent it from boiling (via an air vent valve that releases water and lets air in simultaneously), and (3) it absorbs any air that may get into the system. (Air causes the pipes to bang and may interrupt water circulation.)

Baseboard Heaters The baseboard heater is widely used. It consists of $\frac{1}{2}''$ diameter copper tubing with many fins welded to it. The fins cause heat to radiate quickly because of their large surface area. Baseboard heaters are installed near the floor along the exterior walls. Transferring the heat near the floor reduces the temperature differ-

ential between the floor and the ceiling to only about 3°F.

The pipe and fins are enclosed in sheet steel with baked enamel finish. The enclosure has no bottom and has two openings along its top in order to allow the air to circulate. The room air enters the enclosure through the bottom, moves around the pipe and fins where it gets heated, then exits through the top openings. It is thus important not to block the bottom opening of the baseboard heater by carpeting or flooring, otherwise the air will not circulate. At least $1\frac{1}{2}''$ clearance should be maintained between the bottom of the enclosure and the top of the finish flooring. Baseboard heaters are about nine inches high and three and half inches wide, and therefore they take up a negligible amount of space in the room.

Radiators In areas that do not have much wallspace, such as bathrooms or entrance halls, baseboard heaters may not be sufficient to provide adequate heating, and radiators may be required. Modern radiators are installed partially set into the walls with less than three inches projecting out of the wall. This reduces their interference with the traffic and furniture of the room to a minimum. They also have sleek and attractive designs with covers made of sheet steel with baked enamel finish.

Hot Water Heating Installation A hot water heating system is installed in two distinct stages: (1) rough heat, and (2) finish heat.

Rough Heat. The rough heat starts after the rough framing and roofing are completed. It involves the installation of the supply and return pipes in the walls and floors. The plumber drills holes in the plates, joists, fire blocks, and subflooring to get the pipes through. After the rough heat is completed, the plumber invites the building inspector to approve the work. The plumber then plugs the pipes with paper to prevent debris from getting into them.

Finish Heat. The finish heat starts after the drywall boards are installed. At this stage, the boiler, circulator pumps, baseboard heaters, or radiators are installed, and the boiler is put into operation. The electrical contractor should work closely with the plumber in wiring the boiler, circulator pumps, and thermostats. After the finish heat is completed, the plumber invites the building inspector once more to inspect and approve the work.

Hot Air Heating System

A hot air heating system consists of a **heating furnace,** a **blower,** a **plenum,** a **supply duct, registers,** a **return duct,** a **return,** an **air filter,** and a **humidifier** (Figure 28.2). Each system serves **one** heating zone only.

The heating cycle starts with a signal from the thermostat that switches on the heating furnace and the blower. The blower forces the return air through the heating furnace where it gets heated. The hot air moves into the plenum and

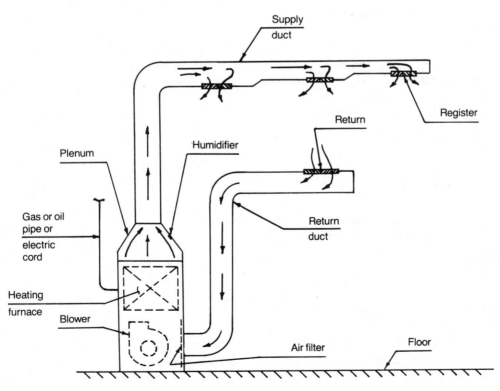

Figure 28.2 Hot air heating system.

the supply duct where it is discharged into the heated area through the registers. The hot air loses some of its heat in raising the temperature of the heated area then returns through the return duct to the blower. This cycle is repeated until the temperature of the heated area reaches the preset temperature of the thermostat, which sends a signal that switches the system off.

Heating Furnace   The heating furnace may be gas, oil, or electric. It heats the air before it gets into the plenum and the supply duct. Gas and oil furnaces have exhaust stacks to discharge the exhaust gases to the outside.

Blower   The blower is located forward of the heating furnace. It is the unit that causes the hot air to circulate. It forces the return air through the heating unit, and thence to the plenum and supply duct. In the meantime, it creates negative pressure in the return duct, which causes the air to be sucked into it through the return. A filter is installed at the end of the return duct in order to retain the dust and prevent it from getting into the system.

Plenum   The plenum is a trapezoid-shaped duct that connects the wide outlet of the heating furnace to the narrower suply duct. It is designed to make the transition as smooth as possible.

Supply Duct and Registers   The supply duct carries the hot air to the heating zone where it releases a portion of the air through each register. The velocity of the air in the duct should be constant. This is accom-plished by reducing the size of the duct after each register (Figure 28.2).

The register consists of a metal cover with louvers that direct the outlet air left and right. Behind the louvers is a damper that can be operated manually to direct the outlet air up and down or to close the register. It is preferable to locate the registers in the floor or in the walls very close to the floor in order to provide uniform heating throughout the area.

Return Duct and Return   The return duct moves the air back to the blower. The air gets into the duct through the return, which consists of a metal cover with louvers but without a damper, so that it is always open. Care must be taken not to block the return with furniture. Usually, there is one return for each heating system. It is preferable to locate the return in the ceiling or in the wall close to the ceiling to allow the hot air to circulate uniformly.

Air Humidifiers   Hot air has low relative humidity, which causes irritation to the human skin. Therefore, hot air heating systems include a power air humidifier that injects a mist into the air flow. The humidifier may be located in the plenum or in the supply duct close to the plenum. The level of humidity is controlled by a humidistat, which switches the humidifier on and off to keep the humidity within preset levels.

Hot Air Heating Installation   A hot air heating system is installed in two stages: (1) **rough heat,** and (2) **finish heat.**

Rough Heat. The rough heat includes the supply and return ducts. It should be installed after the rough framing and roofing are completed. The rough carpenter must make room in the walls and floors to accommodate the ducts. The rough heat should be inspected and approved before the drywall is installed.

Finish Heat. The finish heat includes the rest of the equipment. Installation begins after the drywall is completed.

Central Air Conditioning

A central air conditioning system is similar to central air heating in many respects. The blower, supply duct, registers, return duct, and return are all similar, if not identical, in both systems. The main difference, of course, is that central air conditioning has a cooling unit.

Cooling Unit The cooling unit consists of two parts: one outdoors and one indoors (Figure 28.3). The outdoor part includes a **compressor,** a **condenser** and a **fan.** It is usually installed at the northern side of the house to avoid the direct sun. The indoor part includes an **evaporator coil,** a **blower** and a **condensate drain.** The two parts are interconnected by two **refrigerant copper pipes** that circulate the refrigerant in between them.

The cooling cycle starts with a signal from the thermostat that operates the system. The refrigerant enters the compressor in the form of low pressure gas and leaves it in the form of high pressure gas to the condenser. The condenser causes the refrigerant gas to become liquid and lose heat in the process. The lost heat is forced by the fan to dissipate in the outside air. The refrigerant leaves the condenser in the form of high pressure liquid and goes

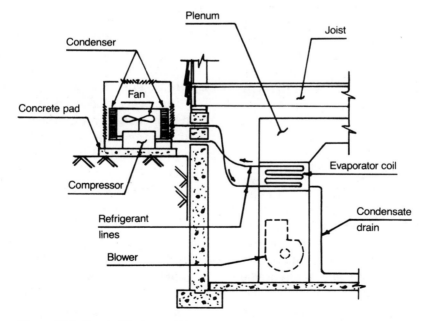

Figure 28.3 Air conditioning system.

to the indoor evaporator coil through one of the refrigerant copper pipes where it is transformed into gas. In the process, it absorbs heat from the air of the house, which is forced by the blower to circulate around the evaporator coil. The gas goes back to the compressor through the other refrigerant pipe where the cycle is repeated.

Cold air holds less humidity than warm air. The difference in humidity condenses into water that is drained from the system through the condensate drain. When the temperature of the house reaches the preset temperature of the thermostat, the latter sends a signal that turns the system off.

Two-Story House Design In a two-story house such as the Colonial, an air conditioning system is designed for each floor. To reduce the amount of ductwork, and thus the initial cost, the indoor part of the second floor unit may be installed in the attic. It should be suspended from the roof rafters in order to reduce the vibrations. In this design, all the registers and the return are located in the ceiling. One disadvantage of this arrangement is that the attic tends to get hot during the summer, which increases the heating load on the system.

Air Conditioning Tonnage Central air conditioning units are rated in tons. **One ton cooling capacity** is equal to **12,000 BTU/Hour** (BTU is defined in Chapter 31). It is the amount of cooling needed to freeze one short ton (2000 pounds) of water into ice in 24 hours. Freezing one pound of water at 32°F into ice at the same temperature requires 144 BTU of cooling.

Heat Pumps

A heat pump can be used for heating and cooling. During winter, it absorbs heat from the outside air and discharges it inside the house. (Cold air includes a considerable amount of heat but we do not feel it.) In summertime, the system is reversed by a switch valve. This causes the heat pump to absorb heat from the house and discharge it outside. Heat pumps run on electricity.

The initial cost of heat pumps is high. In cold areas, they require a back-up heating system, usually electrical resistance coils. The efficiency of heat pumps drops when the outside temperature goes below 15°F. At below freezing temperatures, defrosting the outdoor unit becomes a problem. Heat pumps may be more economical in moderate climates such as the northwest part of the country.

Solar Heating

Solar heating may be considered in the southern part of the country where the sun is abundant and the temperature is high. The components of a solar heating system are: **collectors,** a **heat transfer fluid,** a **circulator pump,** a **heat storage tank,** and a **back-up heating plant** (Figure 28.4). This is in addition to the heat distributing system which may be forced hot water or forced hot air.

Collector A collector consists of a rectangular panel (Figure 28.4), the bottom of which is covered with insulation to retain the heat inside the panel. On top of the insulation lies a black **absorber plate,** that includes many thin parallel tubes. These tubes are con-

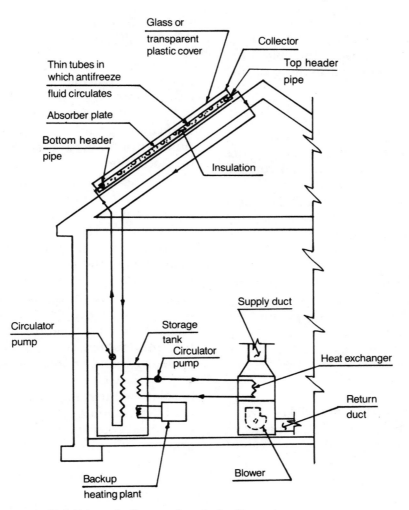

Figure 28.4 Schematic diagram of a solar heating system.

nected to two **header pipes** located at opposite ends of the panel. The panel has a glass or transparent plastic cover that allows the solar radiation into the panel and in the meantime impedes the infrared radiation from the panel. This causes the temperature inside the panel to rise above the ambient temperature.

The collectors must face the sun, which is south. For maximum efficiency, the angle of tilt of the panels should be bigger than the latitude. In practice, panels are usually installed on the side of the roof, which makes their angle of tilt equal to the pitch of the roof.

Heat Transfer Fluid The heat transfer fluid is usually an **antifreeze water solution.** When the sun rises, a circulator pump forces the solution into the collector's thin tubes via the bottom header pipe where it gets heated by the sun. The heated solution moves via the top header pipe to the storage tank where it discharges some of its heat, then goes back to the collectors. This cycle is repeated as long as the sun is shining.

Heat Storage Tank   Solar heating requires a sizeable storage tank to store the collected heat of the sun for use during the night or cloudy days. The tank must be well insulated. The **storage medium** inside the tank may be **water, gravel** or **stones.**

Back-Up Heating Plant   A back-up heating plant is required to fill the gap between the solar heat supply and the household demand. Such a back-up plant may be gas, oil, electric, or a heat pump.

Solar Water Heating   Solar water heating is the most feasible means of harnessing the heat of the sun. It ought to be considered in the Sunbelt.

Contracting for Heating and Air Conditioning

Hot water heating is contracted to a plumber, usually labor and material on a fixed price basis. The price should be based on the design drawings and your particular specifications, such as the number of heating zones, or the use of clock thermostats. The contractor should specify the make and capacity of the boiler. It is preferable to assign both hot water heating and plumbing to one contractor.

Hot air heating should be assigned to an air conditioning contractor. If the house includes both hot air heating and air conditioning, then obviously both have to be contracted to one person. The contract should include the make and tonnage of the cooling units.

Solar heating should be assigned to a specialized contractor. Before signing a solar heating system contract, the contractor should give you the names of several references and you should ask them about the cost and performance of their systems.

CHAPTER 29

Plumbing

Plumbing should be contracted before construction starts, because the plumber can provide the site with a source of water that will be needed during construction. Usually, the trench of the sewage and water pipes is excavated at the same time as the house. By then, the plumber should be prepared to lay the water and sewage pipes and have them inspected in order to have the trench backfilled.

House plumbing consists of three separate pipe systems: (1) cold water, (2) hot water, and (3) drainage. Plumbing also includes the water heater.

Cold and Hot Water Systems

Both cold and hot water systems operate under high pressure. Therefore, the pipes can run up and down without having any adverse effect on their performance. Also, their diameter can be small. The flow of water is completely controlled by valves.

As soon as the main water line enters the house, it is connected to a meter (municipal water supply only) that registers the amount of water consumed.

Next to the meter is a shutoff valve that stops the flow of water to the entire house in case of emergency. The main water line then splits into two lines: one becomes the cold water line and the other goes to the water heater to be heated and becomes the hot water line. The cold and hot water lines run parallel to each other until they reach the valves of the faucets and the appliances where cold and hot water are mixed during usage.

The size and material of hot and cold water pipes are determined by the local building codes and the size of the house. Usually, the diameter of the main lines and the branches should not be less than $1\frac{1}{4}''$ and $\frac{5}{8}''$, respectively. Water pipes may be made of galvanized steel, copper, or plastic, depending on the local building code.

Drainage System

A drainage system operates under gravity. Thus, horizontal pipes have to be pitched toward their points of discharge by about $\frac{1}{4}''$ per foot. Also, they have to be wide in diameter. The horizontal drainage pipes are called

branches; the vertical pipes are called **stacks;** the stacks that carry refuse are called **soil stacks.**

Sewage develops harmful gases. These gases must be vented to the atmosphere by means of vertical stack vents that penetrate the roof (Figure 29.1). Sewage gases, and insects, are prevented from entering into the house by **traps** located at the discharge of each plumbing fixture (Figure 29.1).

Above ground drainage pipes may be made of cast iron, copper, or plastic, depending on the local codes. All underground drainage pipes are made of cast iron. The diameter of the drainage

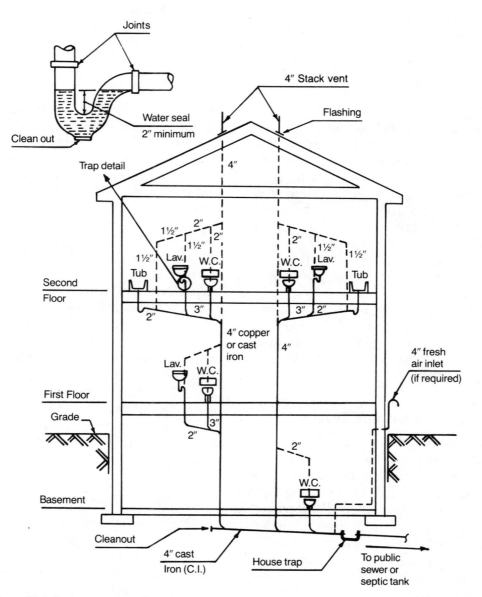

Figure 29.1 Drainage system diagram.

pipes varies from $1\frac{1}{2}''$ for small ventilation pipes to $4''$ for soil stacks. The design drawings should include a drainage system diagram similar to the example shown in Figure 29.1. This diagram includes a layout of the plumbing fixtures, the diameter and material of the drainage and ventilation pipes, and the location of the traps.

Plumbing Installation

Plumbing is installed in two distinct stages: (1) rough plumbing and (2) finish plumbing which includes the water heater and plumbing fixtures.

Rough Plumbing Rough plumbing is the installation of all the pipes, fittings, and traps of the plumbing system. It begins after the wood frame is completed. Most of the rough plumbing is installed inside the walls and ceilings. First, the plumber prepares a detailed drawing showing the diameters, lengths, and material of all the pipes and fittings based on the drawings and field measurements. The plumber may require some alteration in the wood framing in order to get the pipes and fittings through. The rough carpenter should cooperate fully with the plumber. However, you are expected to compensate the carpenter for any extra work.

It is to be noted that the stack walls in which the pipes and soil stacks are installed must not be of the bearing type, meaning that the ceiling joists should not bear on them. They should be wide enough to accommodate the soil stacks, or else boxes have to be built around the stacks.

House Trap Many local codes require that a house trap be installed at a point just before the drainage pipe leaves the house (Figure 29.1). Its purpose is to prevent the gases and insects of the public sewer from filtering into the house.

Testing Most local building codes require that the rough drainage system be **pressure tested** before the drywall is installed. An extension pipe is added to the top of the vent stack to a height of at least 10 feet above the part to be tested. All the open ends of the pipes of the tested section must be plugged or capped. Water is then poured in the top of the extension pipe to the level required by the test. The section passes the test if the level of the water in the extension pipe does not drop within 15 minutes. If the level of water drops, it means that there is a leak that must be fixed. The town inspector must approve the testing. When testing is completed, the plumber covers the openings of all the pipes to prevent construction debris from getting into them.

Water Heater

Most water heaters consist of a heating furnace and a storage tank. The capacity of the storage tank varies between 40 and 80 gallons. The capacity of the water heater must be stated in the plumbing contract. The source of heat may be gas, oil, or electricity. Virtually all water heaters that are being sold today are well insulated. The tank, including the insulation, is encased in a cylindrical cover made of sheet metal with baked enamel finish.

Another type of water heater is the demand type. It has no storage tank. Rather, it consists of a copper coil through which the water to be heated circulates. Upon demand, the water in the coil is heated by either gas or hot water drawn from the boiler. Hot water continues to flow as long as there is demand.

Plumbing Fixtures

Plumbing fixtures are installed in two stages:

1. The bathtubs and showers are installed as soon as the wood frame and roofing are completed. They are heavy and should be left for some time to allow their supporting wood frame to settle and shrink before installing the surrounding ceramic tiles. The plumber must cover the fixtures with glued papers so that they do not get stained, scratched, or cracked.

2. The rest of the fixtures such as the lavatories, toilets, bidets, sinks, faucets, faucet knobs, etc. should be installed just before the house is ready to be occupied, to avoid their being damaged during construction.

After the plumbing fixtures are installed, the system has to be **air tested.** The town inspector or engineer must approve the test.

Contracting for Plumbing

Plumbing may be contracted either labor and materials or labor only. The former is recommended because the responsibility is the plumber's if a fixture is found to be cracked, broken, or defective.

Many local building departments require that the plumber apply for an independent permit and be directly responsible to the building inspector.

You and your family should visit several plumbing fixture showrooms before entering into a plumbing contract. You will enjoy the variety of fixtures in styles, colors, and shapes that dazzle the eyes. Bathtubs, sinks, lavatories, toilets, bidets, showers, faucet knobs, and handles are designed to appeal to various tastes and fit different budgets.

CHAPTER 30

Electrical Wiring and Telephone Line

Electrical wiring is the installation of the circuits that deliver electricity to the fixture outlets, switches, receptacles, house appliances, air conditioning units, fans, etc. The electrical installation should be in accordance with the wiring diagrams that are incorporated in the floor plans. An example of a wiring diagram and the meaning of the most common electrical symbols are shown in Figure 30.1.

You and your family should visit several light fixture showrooms before signing the electrical contract, since you might choose some fixtures that require wiring changes. The showrooms exhibit a fascinating variety of fixtures.

Electrical wiring is installed in two stages: (1) rough wiring, and (2) finish electrical.

The telephone line is wired at the time of the rough wiring, though not by the electrical contractor.

Rough Wiring

Rough wiring begins after the rough heat and rough plumbing are completed. The wiring should be in accordance with the drawings and the electrical codes. Each room should have a minimum number of receptacle outlets depending on its size. The receptacles are installed in the walls a few inches above the baseboard heaters. Switches and receptacle outlets are housed in metal boxes that are fastened to the nearest stud. These boxes also support the cover plate of each item.

The wiring of the house is divided into circuits, each directly connected to the **main circuit breaker.** Separate circuits are provided for the electrical appliances, air conditioning units, outdoor lights, outdoor receptacle outlets, etc. The main circuit breaker includes an on/off switch for the electricity to the entire house.

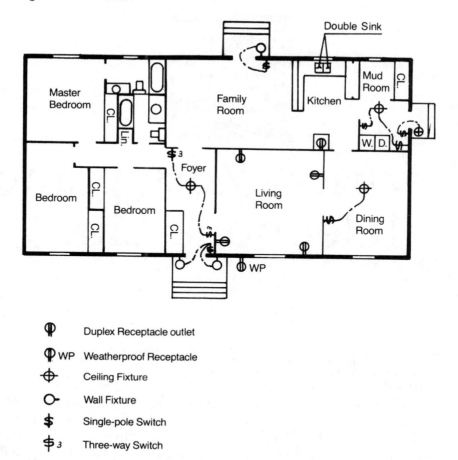

◐	Duplex Receptacle outlet
◐ WP	Weatherproof Receptacle
⊕	Ceiling Fixture
O‑	Wall Fixture
$	Single-pole Switch
$₃	Three-way Switch

Figure 30.1 A floor plan showing part of the electrical wiring diagram.

Finish Electrical

Finish electrical starts after finish heat, air conditioning, and plumbing are installed in order to wire the furnace motors, circulator pumps, dishwasher, central air conditioning units, attic fan, etc. Finish electrical includes the installation of the covers of all the switches and the indoor and outdoor receptacle outlets. The latter should have waterproof covers. It also includes the installation of the light fixtures that you provide. At this stage, it is still possible to add or change the location of some light fixtures or switches.

Finally, the electrician completes the installation of the main circuit breaker and labels each switch to indicate the part of the house it controls. The house is then connected (hooked up) to the electric utility line.

After all the electrical work is completed, the electrical contractor invites the town inspector (without getting you involved) to inspect and approve the wiring. If the inspector is satisfied that the work is done in accordance with the code, a fire underwriters certificate is issued.

Fire Underwriters

The fire underwriters is a certificate that indicates that the electrical work has been done in accordance with the electrical code and has been approved by the building department. It is a very important document that you will need to obtain the **certificate of occupancy** (**c. of o.**) and the last installment of the construction loan. Many lenders require a copy of this certificate before approving a long term mortgage loan.

The fire underwriters certificate includes the number of fixture outlets, receptacles, and switches; the horsepower of the exhaust fans; the power of the dishwasher; and the electric current of the house in amperes.

Contracting for Electrical Wiring

Electrical wiring must be assigned to a licensed and competent contractor who is familiar with the electrical code and the local regulations. Many localities require that the electrical contractor apply for an independent permit and be directly accountable to the town's electrical inspector. The contract should be on a lump sum basis. Choose an electrical contractor as early as possible because he can provide electric power for the carpenter's tools.

Telephone Line

The telephone line is usually wired by the telephone company. At the outset you should advise the company as to whether you intend to use its telephone sets or buy your own. Designer telephones are becoming popular and many homeowners buy them both for their styling and to save the monthly rental fee. The telephone company needs to know the number of telephone outlets, the number of telephone sets, and their code numbers to ensure that the sets you install are compatible with its system. This information is also used to determine your monthly charges.

The telephone wiring is usually connected to the nearest telephone post by a heavy waterproof wire buried in the ground. This wire is provided by the telephone company, but it is your responsibility to dig and backfill the trench in which the wire is laid. In some areas, the telephone wire is suspended on above-ground poles.

CHAPTER 31

Insulation

*T*hermal insulation is the most effective means of conserving energy and reducing its costs. It is accomplished by enveloping the house with a layer of insulating material. **State energy conservation codes** establish the minimum R-Values of the insulation of different parts of the exterior of the house (walls, ceilings, foundations, windows, etc.).

R-Value

R-value stands for the material's thermal resistance to heat flow, that is, heat loss during winter and heat gain during summer. The higher the R-value, the lower the heat loss or gain through the material. In numerical terms,

$$R = \frac{1}{U}$$

where **U** is the heat flow in **BTU** through one square foot of the material per one degree Fahrenheit difference in temperature per hour. One **BTU** (**British Thermal Unit**) is the amount of heat required to raise the temperature of one pound of water one degree Fahrenheit.

If exterior wall insulation is rated R–13, then its U value is equal to

$$\frac{1}{13} = 0.077$$

This means that the heat flow through this material is 0.077 BTU per square foot, per one degree Fahrenheit difference between the inside and outside temperatures, per hour. If the outside temperature is 10°F and the inside temperature is 68°F, the difference in temperature is 58°F. Therefore, the heat flow through one square foot of this material per hour is $0.077 \times 58 = 4.47$ BTU per square foot per hour. The heat flow is usually written in the following format:

$$4.47 \text{ BTU/ft}^2/\text{hr}$$

If the total area of the exterior walls is equal to 1400 square feet, the total heat flow through the walls is equal to

$$4.47 \times 1400 = 6258 \text{ BTU/hr}$$

Forms of Insulation

Insulation is sold in three forms: (1) batts and blankets, (2) loose fill, and (3) rigid boards. In all types of insulation, the R-value and the thickness of the material are clearly printed either on the vapor barrier or the wrapper.

Batts and Blankets   Batts and blankets are classified as flexible insulation. Both are manufactured of either fiberglass or rock wool. The R-value of either material is about 3.0 per inch thickness. Therefore, to get R–13 and R–19 values, the nominal thicknesses of batts or blankets should be about four and six inches, respectively.

The difference between batts and blankets is that batts are sold in nominal lengths of four or eight feet, to fit perfectly in eight-foot high walls, while blankets are sold in rolls to be cut on site. They are both pre-cut to 15- and 23-inch wide strips to fit tightly between wood framing spaced at 16 and 24 inches o.c., respectively. Both are sold either with or without vapor barrier attached to one face. The attached vapor barrier has a one-inch wide flange folded along each side of the strips for stapling the insulation to the wood frame.

Loose Fill   Loose fill insulation is made of vermiculite, cellulose, or other materials. It is sold in bags mostly in granular form. Loose fill is either poured in open areas such as the floor of the attic, or blown into closed areas. The R-value per one-inch thickness of the material should be clearly marked on the bags.

Rigid Board   Rigid board insulation is manufactured of polystyrene, polyurethane, or other material. It is sold in 24- or 48-inch wide boards. Most rigid board insulation does not require vapor barrier. One big disadvantage of several types of this insulation is that they are not fire resistant. Therefore, they have to be covered with stucco or drywall, depending upon where they are installed.

Vapor Barrier

A vapor barrier must be installed on the warm side of the insulation (facing the heated area). Its purpose is to prevent the water vapor created in the house during bathing, cooking, laundering, etc., from getting into the insulation. During cold weather, this water vapor penetrates the walls and ceiling of the house, and, without a vapor barrier, will reach and condensate into the insulation causing it to lose its insulating characteristics.

Air Barrier

An air barrier may be installed over the exterior sheathing of the house as an extra insulating measure. It prevents infiltration of cold air into the house, but allows any water vapor that may get into the insulation to pass through it to the outside. To be fully effective, the air barrier must fit tightly around the windows, exterior doors, and other openings.

Installing Insulation

Installing insulation is simple enough that you may do it yourself, if you have the time and energy. But it has to be done with care. Improper installation of vapor barrier, or gaps left between the insulation and the adjacent house framing or around the windows and exterior doors markedly reduces the effectiveness of the insulation.

Walls Walls may be insulated by either batts or blankets with or without attached vapor barrier. Insulation should fit tightly against the studs and the top and sole plates. If batts or blankets with attached vapor barrier are used, the barrier flanges should be stapled to the inner sides of the studs starting from the top going down every eight inches. If batts or blankets without vapor barrier are installed, the entire side of the wall facing the heated area must be covered with a vapor barrier consisting of 4- to 6-mil-thick **polyethylene** sheets.

It is very important to install the insulation around water pipes and wall radiators to reduce heat loss and prevent the water in the pipes from freezing. It is equally important to stuff the area around window and door framing with loose insulation. Any rips in the vapor barrier must be patched with tape.

Attics The layout of the attic insulation depends on whether it is heated or unheated. Figure 31.1 shows layouts for heated and unheated attics. In all cases, insulation must not cover the eave vents, or block the passage of air ventilation. A baffle should be installed at each eave vent (Figure 31.2) to retain the loose-fill type of insulation. Insulation should be kept three inches away from recessed light fixtures, otherwise they will overheat.

As mentioned earlier, the side of the insulation facing the heated area must

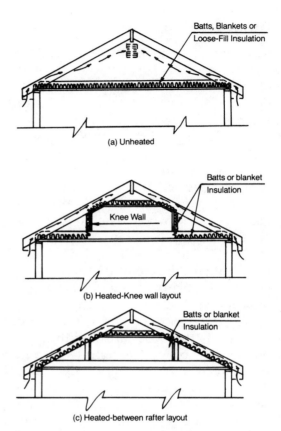

Figure 31.1 Insulation layout for attics.

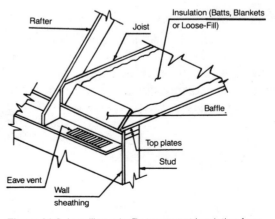

Figure 31.2 Installing a baffle to prevent insulation from clogging eave vents.

have a vapor barrier. If the insulation consists of two layers, the second (top) layer must not have a vapor barrier.

The insulation of unheated attics is laid between the ceiling joints. It must cover the top plates of the exterior walls. Workers should not walk on the insulation.

There are two layouts for installing the insulation of heated attics. In one layout, the insulation of the floor of the attic is installed starting at the eave and ending at the knee wall. Afterward, the knee wall and the attic ceiling are installed (Figure 31.1b). In the other layout, the insulation is installed between the rafters starting at the eaves and going up to and over the attic ceiling (Figure 31.1c). This layout uses less insulation, involves less work, and insulates better because it has less exposed surface.

Cathedral Ceiling There are two methods of insulating cathedral ceilings depending on whether the interior of the ceiling is covered or exposed:

1. For covered ceilings, batts or blankets are laid between the rafters (Figure 31.3a). The vapor barrier should face the heated area (inside the house). Care must be taken not to obstruct air ventilation.

2. For exposed ceilings, rigid insulation is laid on top of the roof decking (Figure

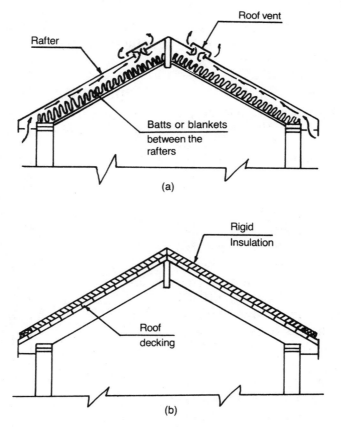

Figure 31.3 Insulating a cathedral ceiling.

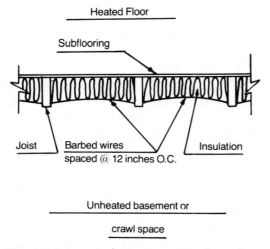

Figure 31.4 Insulation of floors over unheated spaces.

31.3b). It is important to follow the manu-facturer's recommendations regarding exposed installation.

Floors over Unheated Areas Floors over unheated basements or crawl spaces must be insulated to reduce heat loss. Batts and blankets with vapor barrier facing up are installed between the joists. Insulation is secured in place by either barbed wire (Figure 31.4) spaced at 12 inches, or by stringing a wire or fishing line between the bottom of the joists.

Basement Walls Before installing in-sulation in basement walls, you must be sure that the walls are dry. Moisture can render fiberglass and rock wool insulation useless. The best time to check the walls is after heavy rainfall. If water seepage is significant, the ex-terior of the walls must be repaired before the insulation is installed. But if the seepage is through hair cracks in poured concrete walls, the cracks may heal after a few rainfalls. Rain water carries with it very fine soil particles that may seal the cracks. Therefore,

you must hold off installation for a few months until this process takes its course.

Once you have ascertained that moisture is not a problem, the walls should be treated with a waterproofing coat that extends up to the grade ele-vation (Figure 31.5). The carpenter may then proceed with constructing the sole plate, studs, and top plate. Batts or blankets are then installed between the studs. Usually, four-inch thick batts or blankets are adequate for cold climates. The inside face of the insulation must have a vapor barrier.

Basement walls may be insulated with rigid boards instead of batts or blankets. The boards are attached to the walls by either an adhesive or by using **furring strips and fasteners.** If insulation is not fire resistant, it must be covered with drywall sheets or an-other fire resistant material.

Contracting for Insulation

You may contract the insulation ma-terial and labor, or buy the material

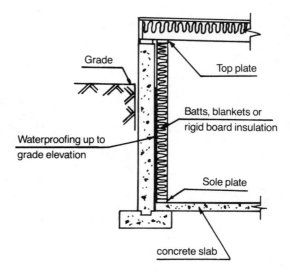

Figure 31.5 Insulating basement walls.

and contract the labor only. The amount in square feet and the R-value of the insulation for each part of the exterior of the house should be included in the list of materials enclosed with the design drawings. In general, the cost of labor is about 80 percent of the cost of the materials. A laborer can install about 100 square feet of batts or blankets per hour.

CHAPTER 32

Stairs

S tairs are the series of steps or flights that lead from one floor to another. The principal stair components are shown in Figure 32.1.

Stair Components

Tread. The horizontal face of a step that users step on when ascending or descending from one floor to another.
Nosing. The extension of a tread beyond the face of the riser below. Its edge is usually rounded for safety and appearance.
Riser. The vertical face of a step.
Flight. A series of steps leading straight from one landing to another.
Landing. A level platform located at the beginning and end of each flight.
Run. The horizontal projection of a flight.
Rise. The vertical projection of a flight.
Stringer. The inclined member that supports the steps of main stairs. Open stringers have their top cut following the lines of the treads and risers. Closed stringers have parallel top and bottom edges.
Carriage. The rough timber planks that support the steps of service or basement stairs.

Staircase. The area containing the stairs. Also, a flight of stairs including the balusters and handrailing.
Baluster. The vertical members supporting the railing in open stairs.
Newel post. The main post that supports the railing at each end.
Handrail. The top railing that users grasp with the hand while ascending or descending the stairs.

Stair Design

All stairs must be designed for safety and comfort. In any flight, the width of all treads must be equal, and the height of all risers must be equal. If this is not done, users might stumble. Treads have nosings, usually rounded, to increase the area of the treads, and lessen the chance of harm to the front of the leg if the nosing is accidentally kicked while ascending the stairs.

The width of the tread plus the height of the riser should be neither too long nor too short for maximum safety and comfort. To attain a balanced design, (1) the tread width multiplied by the riser height should be between 72 and 75, and (2) the tread width plus twice the riser height should be between 24

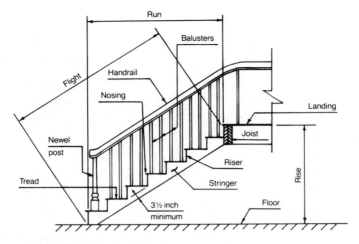

Figure 32.1 A typical stair.

and $25\frac{1}{2}$ inches. According to the first condition, if a riser is 7 inches high, the tread should be about $10\frac{1}{2}$ inches wide ($7 \times 10\frac{1}{2} = 73\frac{1}{2}$). These dimensions satisfy the second condition ($10\frac{1}{2} + 2 \times 7 = 24\frac{1}{2}$).

Stairs should have a minimum headroom of 6'-8" for main stairs and 6'-4" for service or basement stairs. Open stairs must have handrails and balusters along their sides. The handrails provide the users with something to hold on to, and the balusters prevent children from falling off the sides of the stairs. The handrails of closed stairs may be fastened to the walls by metal brackets, in place of the balusters.

Types of Stairs

Two types of stairs are used most often in house construction: (1) main stairs, which lead from one floor to another, and (2) service stairs, which lead to the basement or garage. In addition to these, some houses may have folding attic stairs.

Main Stairs The main stairs are designed with emphasis on comfort and appearance. Comfort is provided by using low risers, wide treads, and wide stairs. Generally, the height of the risers should not exceed $7\frac{1}{2}''$, and the width of the stairs should not be less than 3'-6".

For attractive appearance, stair manufacturers use expensive hardwood such as oak, birch, or maple; grooved-type connection between the tread and riser (Figure 32.2a); and decorative balusters, newel posts, and handrailing. To reduce costs, the risers, which are not subjected to wear, may be made of less expensive wood such as Douglas fir or southern pine.

Service Stairs Service or basement stairs are designed with emphasis on economy. They are made entirely of the less expensive woods such as Douglas fir or southern pine. The height of the risers can be 8 inches or slightly more; and the width of the stairs 3 feet or less. The stairs may be supported by simple carriages consisting of rough wood planks. The connection between the riser and tread is simple (Figure 32.2b). Some stairs may consist of $1\frac{1}{2}''$-thick plank treads, with no risers. But

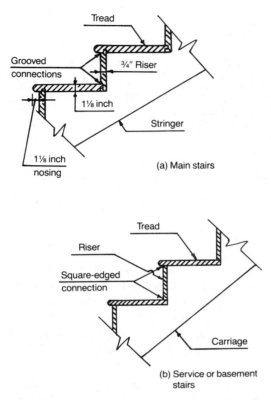

Figure 32.2 Stair details.

Contracting for Stairs

Stairs are usually assigned to a workshop after the rough framing and roofing are completed. The workshop arranges for a trained representative to meet with you at the house. Make sure that the rough carpenter is also present.

The workshop representative starts by measuring the exact height of each story and the staircase opening in each floor, and suggesting the best layout regarding the width of the treads, the height of the risers and where the stairs should begin and end. All this should be done in cooperation with the rough carpenter, and with your consent. In the meantime, the representative will show you catalogues and samples of various stair designs and styles to choose from. Get a range of ideas and prices from several sources before signing a contract. You will be impressed by the variety of the shapes and styles of stairs, balusters, newel posts, and handrails that each workshop has to offer.

Stairs are shipped to your house in the form of flights consisting of stringers, treads, and risers put together and wrapped for their protection. They are installed by the rough carpenter who also builds the landings that serve the flights. Stair installation should be delayed as much as possible to reduce their exposure to construction traffic. Delivery and installation of balusters, newel posts, and the handrailing should be put off until the house is about to be occupied, to avoid damage during construction.

this design is not recommended since it is deficient in both strength and appearance.

Attic Folding Stairs   Attic folding stairs are used where there is no room for fixed stairs, such as in the garage. They are housed in the attic until needed. They slide down through a hole in the ceiling by means of a rope. The floor and joists of attics served by folding stairs must be designed to support partial floor loadings.

Fireplace and Chimney

A fireplace is mostly a decorative item. It adds considerably to the attractiveness and value of the house even though its heating efficiency is very low. Having a fire going on a cold night creates a cozy atmosphere of relaxation and comfort.

Fireplaces are optional; low-cost houses do not have any, moderately priced houses have one, and expensive houses may have several.

Conventional and Prefabricated Fireplaces

A conventional fireplace is expensive, in itself worth a few thousand dollars. It may be built of bricks or stones and may have a decorative mantel around its front opening. It exhausts its smoke into the atmosphere through a flue pipe embedded in a chimney.

A prefabricated fireplace is inexpensive and lightweight. It consists of a compact metal firebox, ash tray, and an insulated free standing sheet metal chimney. It can be installed in any area where it is feasible to cut a hole in the ceiling to allow the chimney to pass through. Some local building codes impose restrictions on prefabricated fireplace installation.

Fireplace Lighting

Most fireplaces burn wood logs. However, some homeowners prefer gas burning units because they are ash and smoke free. They consist of gas tubes hidden in imitation wood logs made of fire resistant material such as ceramic or clay. The aesthetic disadvantage of a gas fire is that it does not glow like a wood fire.

Electrically lit fireplaces, on the other hand, do not burn any fuel, and thus do not require a chimney. They consist of imitation wood logs with rotating lights that create the impression of a glowing fire, but without fire or heat.

Fireplace Components

A fireplace consists of a firebox, a front hearth, air vent, ash pit, throat, damper, smoke chamber, and a smoke shelf (Figure 33.1).

Firebox A firebox is the chamber in which combustion takes place. It must

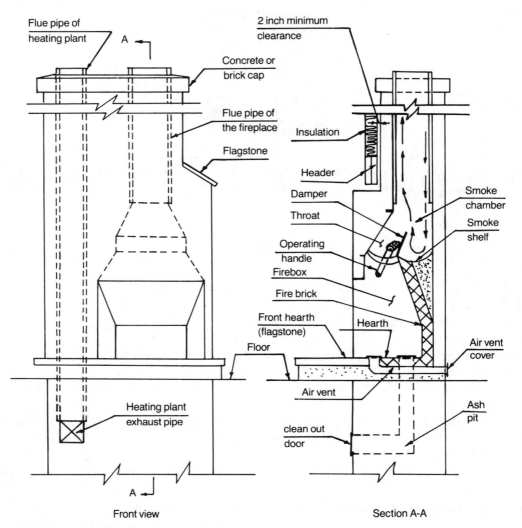

Figure 33.1 Fireplace and chimney.

be lined with fire resistant material such as fire bricks. The bottom of the firebox is called **the hearth**. For safety, the elevation of the hearth should be at least 8 inches higher than the elevation of the floor of the room.

The back of the firebox should be narrower than the front in order to expedite the discharge of the smoke and combustion gases and to reflect more heat into the room.

The size of the front opening of the firebox should meet two criteria: (1) its height should be about three quarters of its width, and (2) its height should be about 1.5 times the depth of the firebox.

Front Hearth The front hearth is an extension of the hearth. It extends into the room by about 18″. Sidewise, it extends beyond each side of the firebox opening by 12", minimum. Its function is to protect the floor of the room from flying sparks.

Air Vent Many building codes require that the firebox be connected to the outside with an air vent (Figure 33.1). Its function is to draw in the air required for combustion from outside instead of consuming the heated air of the room. The exterior opening of the air vent must be covered with a screen to prevent rodents and insects from getting into the house.

Ash Pit The ash pit is a small chamber located in the bottom of the firebox. It collects the ash resulting from the fire. Every so often, the ash pit is emptied through a clean-out door located either in the basement or outside.

Throat The throat is the narrow exit located at the top of the firebox through which exhaust gases and smoke pass to the smoke chamber, then to the flue pipe, and finally to the outside. It is called the throat because its size is much smaller than that of the firebox. For good draft, the area of the throat should be about 1.33 times the area of the flue pipe.

Damper The damper is located within the throat. Its function is to control the draft. It should be fully open for a strong fire, and partially open for a small fire. During cold weather, the damper should be closed when the fireplace is not lit to prevent the heated air of the room from escaping to the outside. The damper should also be closed during warm weather to pre-

vent rodents and insects from getting into the house.

Smoke Chamber The smoke chamber is located above the throat. It provides a smooth transition between the throat and the flue pipe.

Smoke Shelf The smoke shelf is located at the bottom of the smoke chamber along the top of the throat. It is curved upward in order to deflect the cold air that flows down through the flue pipe. This action prevents the cold air from getting into the firebox and disrupting the fire.

Prefabricated Fireplace Form

A prefabricated fireplace form is an assembly made of steel. It consists of a firebox, throat, damper, smoke chamber, and smoke shelf. It facilitates the construction of the fireplace and ensures that the firebox will have the correct proportions. Fireplace forms are sold in different sizes at building material suppliers and fireplace shops. The size you choose must be compatible with the size of the fireplace foundations.

Heat Efficiency

To increase the efficiency of the fireplace, a built-in heating duct may be installed at each side of the firebox. The room air enters the ducts through the lower openings, gets heated in the duct, then exits through the higher openings. These heating ducts may be prefabricated or built into the bricks.

Efficiency may also be increased by installing a fire resistant glass door on the fireplace front opening, to prevent the heated air of the room from entering the firebox and escaping to the outside.

Chimney

A chimney has two main parts: (1) flue pipes, and (2) the brick structure surrounding the flue pipes.

Flue Pipes The flue pipes are used to line the hollow shafts through which the smoke and exhaust gases of the fireplace and heating plants are discharged to the atmosphere (these shafts are called flues). They consist of square, rectangular, or circular tubes. They are usually made of vitrified clay, which is good at resisting both heat and acids (combustion gases produce acids when they get in contact with moisture). Their inner surface is smooth, which helps the draft. The tube lengths vary from one to two feet; the fewer the joints, the more air and smoke tightness. The size of the flue pipes may be 8″ × 12″ or 12″ × 12″ for fireplaces, and 8″ × 8″ or 8″ diameter for heating plants (all nominal sizes).

Masonry Structure The masonry (brick or stone) structure surrounding the flue pipes has three basic functions: (1) to support the flue pipes and protect them from forces of nature such as wind and earthquakes, (2) to protect the wood frame of the house from the fire hazard of the hot gases in the flues, and (3) to keep the exhaust gases hot, which is essential to the draft. Each

flue pipe must be surrounded by at least 4″-thick masonry on all sides.

For further protection, the outer face of the chimney should not be closer than 2″ from the nearest wood framing.

Chimney's Height The minimum height of the chimney depends on its location with respect to the ridge of the roof. If the chimney is located at or close to the ridge, its top should be 2′-0″ higher than the ridge. But if the ridge is too far from the chimney, the elevation of the roof at a horizontal distance from the chimney equal to 10′-0″ is used as a base line. The height of the chimney should be 2′-0″ above the base line.

Contracting for a Fireplace and Chimney

The fireplace and chimney may be contracted either material and labor or labor only. The mason must be familiar with the building codes regarding the height of the chimney and the air vent. If the exterior siding includes brick or stone veneer, the contractor who builds the veneer should also build the fireplace and chimney.

It is extremely important that the mason be competent since there are many things that can go wrong in fireplace and chimney construction. Your contract should include a clause that the contractor must clean off the brick before the mortar hardens in accordance with the recommendations of the brick manufacturer.

CHAPTER 34

Drywall and Plastering

Drywall and plastering are used to cover the interior walls and ceilings. They also serve as a base for painting, wall papering or any other type of interior finish.

Drywall

Drywall boards, also called **gypsum board, plasterboard,** or **sheetrock,** are very popular because they are quick and easy to install, economical, sound and fire proof, and less messy than plastering. They consist of a gypsum core faced with paper on both sides. The boards are 4′ wide, 8′ to 16′ long, and $\frac{3}{8}$″ to $\frac{5}{8}$″ thick. They are tapered along their edges to accommodate the paper tape that covers the joints.

In many areas, the building codes establish the minimum drywall thicknesses, usually $\frac{1}{2}$″ for the walls and ceilings of the house and $\frac{5}{8}$″ for the walls and ceilings of the garage. Building codes also require that the wood framing in the vicinity of the boiler must be covered with $\frac{5}{8}$″-thick drywall extending 3′ beyond the boiler's boundaries.

Drywall Installation

Drywall is installed after the plumbing pipes, heating and air conditioning pipes and ducts, electrical wiring, telephone line, and insulation are installed and, except for the telephone line, inspected.

In small rooms, the drywall is installed vertically using 4′ × 8′ boards. In big rooms, the drywall is installed horizontally using long boards to reduce the amount of joints. In bathrooms, it is a good idea not to install drywall on the walls surrounding the bathtub and shower heads. These areas should be covered with **wonderboards,** which are waterproof.

Before installation begins, the drywall installer should lay out the ceiling and walls of each room using boards of different lengths to minimize board cuttings. The joints should be staggered (Figure 34.1). A little space should be provided between adjacent boards to allow for drywall expansion due to humidity and temperature changes. Joints over the corners of windows and doors should be avoided

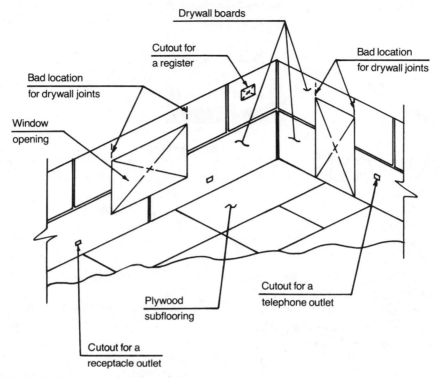

Figure 34.1 Horizontally laid drywall.

since this is where cracking due to wood shrinkage occurs the most.

The ceilings are installed first, followed by the walls. The drywall installer must make all the necessary cuts for the electrical outlets, telephone outlets, plumbing pipes, air duct registers and returns, and any other installations such as an intercom, central vacuum, etc.

Methods of Fastening Drywall Drywall may be fastened to the wood frame by nails, screws, or an adhesive.

Nails are the most common fasteners. Their disadvantage is that they tend to pop out after a while when the lumber to which they are fastened loses some of its moisture and shrinks. To reduce nail popping, the wood frame

should be allowed to stand for some time to dry out before installing the drywall. Another solution is to use **ring-shank** nails, which are shorter than regular nails, but have the same fastening capability.

Nail popping can be reduced significantly by fastening the drywall with special Phillips-head **screws**. However, they cost more than nails and require electricity to power the screw driver.

Drywall may also be fastened to the wood framing with an adhesive. The application of the adhesive should be in accordance with the manufacturer's recommendations.

Corner Beads Corner beads consist of small angles made of perforated metal or expanded metal lath. They are installed on all exterior corners to shape and reinforce them. The beads must be installed plumb. Once installed, care must be taken not to dent or twist them.

Cleaning the House Drywall installers leave behind a lot of board cuttings that must be removed before taping starts. Most likely, this will be your responsibility since most contractors exclude cleaning the house from their contracts. Their workers are highly skilled and highly paid, and therefore it is not practicable for them to do menial work. Furthermore, a contractor who includes cleaning the house in his price may lose the bid to another one who does not.

Taping

Taping is the covering of drywall joints with paper tape, then covering the tape, the heads of the nails and screws, and the corner beads with several coats of a ready-mix joint compound. The purpose of taping is to transform the drywall boards into one smooth surface that is ready for painting, wall papering or any other type of finish.

Taping begins with the ceiling joints, followed by the wall corners and joints, and finally the corners between the ceiling and the walls. First, the taper spreads a layer of the ready-mix joint compound over the tapered edges of the boards. This layer fills the joints partially and provides a surface to which the paper tape adheres. The taper installs the paper tape over the joints. The tape is about two inches wide and is sold in 250-foot long rolls. Next, the taper covers the tape, the

heads of the nails and screws, and the corner beads with the joint compound, going over the compound with a broad knife to produce a smooth surface. This process is called the first coat. After the joint compound dries, in about 24 hours, the rough areas are sanded.

The next step is to apply the second coat. This coat should extend beyond the first one in order to produce a smoother surface. After waiting another 24 hours and sanding it, the taper applies the final coat.

It is common practice to apply only two coats to the inside of the closets and only one coat to the ceiling and walls of the garage.

During taping, the temperature of the house should be at least 70°F in order for the compound to dry. In winter, the central heating system must be in operation before taping begins. Some people use portable kerosene heaters, but this is not recommended since they are considered to be a fire hazard.

One-Coat Plastering

Some homeowners prefer plaster to drywall. It has a hard plane surface that is free of joints and nail popping. It can also hide the imperfections of the wood frame. An economical way to accomplish this is **one-coat plastering.**

With this method, a type of gypsum board called **blue board** is installed over the wood framing. It is covered with one layer of ready-mix plaster (plaster adheres well to blue boards). Before the plaster dries out, the plasterer finishes the surface according to your specifications.

Contracting for Drywall and Plastering

There are several ways to contract drywall and plastering:

1. To contract the entire job material and labor to one contractor.

2. To contract the material and labor of drywall or blue board installation to one contractor, and the material and labor of taping or plastering to another (board installation, taping, and plastering are different professions).

3. To buy the materials and contract board installation, taping, and plastering either separately or combined.

The first method saves a lot of time and effort. On average, the cost of drywall and tape installation is a three-way split: one third for the drywall boards, one third for the drywall installation, and one third for taping.

It should be emphasized that the quality of taping or plastering has a lasting effect on the appearance of the walls and ceilings of the house. Visible bulging or curling tape, popped-out nail heads or imperfect plastering degrades the value of the house. Thus, taping and plastering must be contracted to professionals with many years of experience.

CHAPTER 35

Underlayment, Interior Doors, and Interior Trim

The installation of the underlayment, interior doors and interior trim starts after the drywall (or plastering) and the primer-sealer is completed, and the floor is broom cleaned. It is recommended to apply the primer-sealer coat of paint before in order to avoid staining the wood work.

The term **interior trim** includes the installation of the interior doors, their **casings, hinges, stops and hardware;** the **window stools, casings, aprons,** and **hardware;** the **closet shelves** and **hang rods;** and the **base molding.**

Underlayment

The underlayment consists of $\frac{1}{2}$"-thick plywood or particlewood panels. It is installed over the subflooring in the areas where wall-to-wall carpeting is to be installed. Its function is to provide the carpeting with a smooth, clean, and rigid base.

The installation of the underlayment panels is similar to that of plywood subflooring. Care must be taken that the joints of the underlayment do not coincide with the joints of the subflooring. The underlayment is installed before the interior doors and interior trim.

Interior Doors

Interior doors are available in various types and materials. The types include **flush, sliding,** or **bypass** (for closets), and **folding** (for closets and kitchen). The predominant material for interior door construction is wood. Some folding doors are made of vinyl. The door between the garage and the house and that of the boiler room must be made of fire proof material, such as steel.

Usually, interior doors are sold assembled (**Pre-hung**). The basic assembly includes the door and its frame (**the jambs**). Some assemblies include the casings and door stops. The assembly is braced to prevent it from distorting during transportation.

The height of most interior doors is 80″. Their width starts from 24″ and increases by 2″ increments. The design drawings should include a door schedule indicating the size, type, and material of each door. Furthermore, the design drawings should indicate the direction in which the doors will swing. For uniform appearance, it is recommended that all the doors on one floor be of the same type.

Panel Doors Interior panel doors are similar to the exterior doors (see Chapter 23). The main difference between them is that interior doors are $1\frac{3}{8}$″ thick, while exterior doors are $1\frac{3}{4}$″ thick.

Flush Doors Flush doors may be **hollow-core** or **solid-core**. Most interior doors are of the hollow-core type. They consist of a light wood frame faced on both sides with plywood panels made of a hardwood such as birch, oak, or mahogany. The core is filled with lightweight materials, such as corrugated cardboard, that are glued to the face panels. Solid blockings are installed at one or both sides of the door to accommodate the **mortise** for the lock.

Solid-core doors are different from hollow-core in that their core is made of blocks of wood laid staggered and glued to the face panels and to each other. This makes them very strong. Solid-core doors may be used for bathrooms where there can be a considerable difference in humidity from one side of the door to the other.

Sliding Doors Sliding, or **bypass,** doors are sold in pairs. They are used mainly for closets. They open and close by rolling along an overhead track. Their bottom is guided by plastic or steel angles fastened to the floor. If the floor of the closet is carpeted, the angles have to be installed over wood blocks.

Folding Doors Folding doors are usually **louvered**. They are used for closets and kitchens. They are sold in pairs hinged together. Their top is guided in a track to prevent them from swinging in and out.

Installation of Interior Doors and Interior Trim

In order to protect the finish wood flooring or carpeting, it is recommended that the interior doors and interior trim be installed first. The carpenter should be given the type and thicknesses of the flooring material for each room in advance so that he may raise the doors by that amount plus a little clearance.

Door and Trim The first step is to install each door frame in the corresponding wall opening. The frame is plumbed and leveled using wood wedges or shims between the side jambs and the supporting studs. After ensuring that the door fits in the frame, the joints between the door frame and the surrounding studs and header are covered on both sides with **casings** (Figure 35.1). The corner joints of the casings are usually **mitered**. The next step is to install the door **hinges, lock,**

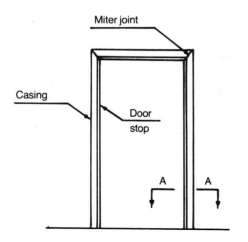

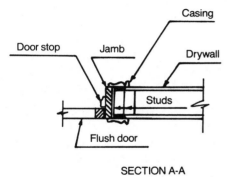

SECTION A-A

Figure 35.1 Interior door trim.

knob, and **stop.** The stop must have full contact with the door when it closes.

Window Trim The first piece to be installed in the window trim is the **stool,** which is the horizontal molding located above the window sill (Figure 35.2). It extends beyond the edges of the window to form a base for the side casings. Next, the side and top casings are installed in a way similar to door casings. Finally, the **apron** is installed below the stool.

If the thickness of the exterior walls

is six inches, an extra two-inch wide trim will have to be installed along the inner edges of the window frame, since most window frames are manufactured to fit into four-inch thick walls.

Closets There are several types of closets. **Clothes closets** have one or two shelves and a hanging rod. **Linen closets** have four or five shelves. **Walk-in closets** have one or two shelves and one hanging rod on each side. The kitchen may have a **pantry closet** that includes four or five shelves or a space for the broom and other cleaning equipment.

The shelves may be made of wood planks, particleboard, or wire mesh with white enamel finish. The mesh is becoming popular because it has the

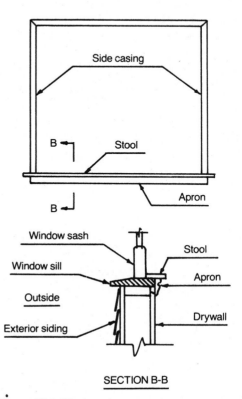

SECTION B-B

Figure 35.2 Window trim.
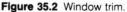

modern look, light, and does not require painting or staining.

Base Molding   The base moldings are used to cover the joints between the wall and the finish wood floor or the plywood underlayment (ceramic and marble floors are furnished with ceramic and marble base molding). They vary in shape and size, and accordingly, in price. Your building material supplier should show you samples of the base moldings he or she carries.

Checking the Trim Work

It is important to check the trim work before paying the last installment to the carpenter. First, open and close the doors several times. The edge of the doors should have continuous contact with the stops. A little force should be required to close the doors so that they do not rattle. Check also that the locks function properly.

The windows must open and close smoothly, and their locks must function properly. It is extremely important to check that the frames of the windows are square by seeing that the bottom and sides of the sash are in full contact with the window sill and jambs.

One way of checking the carpenter's workmanship is to observe the casing's miter joints. If the joints are precise, it is an indication that the carpenter is paying attention to details. But if the joints are open or if their sides do not match, the carpenter is doing a sloppy job.

If the moisture content of the casing is relatively high, the outer edges of the miter joints may open slightly when the lumber dries out. One solution is to let the casing stand indoors for a few days before installation to allow its moisture content to stabilize.

Contracting for Underlayment, Interior Doors, and Interior Trim

These items should be a part of the carpenter's contract. Usually, you buy the plywood, doors, and the trim, and contract labor only. The labor cost depends on the size of the house, the area of the underlayment, and the number of doors and closets, etc., and therefore should be determined by competitive bidding.

CHAPTER 36

Paints and Stains

Paints are used to finish the interior walls and ceilings, windows, doors, trim, exterior siding, decks, and porches; **stains** are used to finish wood doors, windows, trim, decks, and porches.

Exterior and Interior Paints

The main difference between exterior and interior paints is that the exterior contain more **resin** and **pigment,** which increases their durability and weather resistance. Naturally, this makes them more expensive.

The first coat to be applied to new drywall, plaster, or wood should be a **primer-sealer,** which is designed to seal and adhere to new surfaces. After the primer-sealer dries, one or two coats of finish paints are applied.

Paint may be **latex-, alkyd-,** or **oil-based.** These are the basic characteristics of each:

Latex-Base Paints Latex-base paints are thinned with water. They dry quickly making it possible to apply two coats in the same day. They are porous, allowing the moisture to pass through

them, and this minimizes their blistering and peeling. Latex-base paints are popular because they are relatively inexpensive and the touches can be cleaned up with water and soap.

Alkyd-Base Paints Alkyd-base paints are thinned with solvents. They are more durable than latex paints. They also bond better to porous and dusty, but not oily, surfaces. Because of these qualities, alkyd-base are more expensive than comparable latex-base paints.

Oil-Base Paints Oil-base paints are also thinned with solvents. However, they are not popular because they dry slowly making them vulnerable to weather changes and bugs. They also have a strong, undesirable odor. Their biggest advantage is durability.

Paint Lusters Exterior and interior paints are sold in three basic lusters: **flat, semigloss** or **satin,** and **gloss** or **high-gloss.** Flat paints hide little imperfections in the painted surface. On the other hand, satin and gloss paints are more durable and better resist moisture penetration. Their disadvantage is that they do not hide imperfections in the painted surface. They are also more expensive than flat paints.

Enamels   Enamel is a combination of paint and varnish. It dries to a hard smooth semigloss or high-gloss luster. Enamels are used to finish wood.

Exterior and Interior Stains

Exterior and interior stains are used instead of paints to finish wood surfaces. Exterior stains are more durable, weather resistant, and therefore, more expensive than comparable interior stains. The most common types are water stains, oil stains and sealing stains.

Most stains are semitransparent, which allows much of the grain pattern to show through. Stains are superior to paints in that they penetrate the wood, and therefore do not blister or peel.

Wood Preservatives

Wood preservatives repel water and prevent wood rot. They may be either clear or semicolored. The clear types show the natural grain of the wood.

A coat of clear wood preservative may be applied to the wood siding, deck, or porch to protect it until you decide on the stain colors. This coat can last for up to a year.

Colors of Paints and Stains

Each manufacturer has charts showing the colors of its paints or stains. However, the colors in these charts may not look exactly the same on your drywall, plastering, or wood. It is therefore recommended that you try the colors on small areas before painting or staining the entire job. Paint dealers can change the colors by changing the **pigment mix (the formula)**. You should get the formulas of the paints you buy in case you need more of the same in the future.

Varnish

Stained interior doors, windows, and trim usually receive one or two coats of **polyurethane varnish.** It is clear and highly resistant to moisture, scratching, grease, and food stains. Two coats of varnish can last a lifetime. Exterior wood should not be varnished, however.

Choosing Paints and Stains

Paints and stains contribute significantly to the final appearance of your house. Therefore, it is important that you do some planning regarding the right combination of paint and stain types, colors, and luster and by observing the finishes of your neighbors' houses before starting.

In addition to your taste and preference, the type of paint and stain you will choose depends largely on the material to be finished. For your guidance, different materials and how they can best be finished are discussed in the next few paragraphs. It should be emphasized, however, that good results may be obtained by other than the finishes suggested here.

Wood Shingles and Shakes   Shingles and shakes made of cedar or redwood do not need protection. They weather naturally into a dark silvery color. They may be given a coat of clear wood preservative before they are installed. If you desire to color the shingles and shakes, they should be stained rather than painted. Because they are porous, they will soak up a lot of stain. To solve this problem, they should be sealed first with a clear penetrating sealer.

Wood Siding   Wood siding should be stained. If painted, it usually receives a coat of primer-sealer, then one or two coats of an alkyd-base paint with a semigloss or high-gloss luster. Latex paints have a dull finish and this makes them unpopular.

Wood Decks and Porches   It is preferable to stain wood decks and porches using decking stains that resist abrasion. Decks and porches that are constructed of pressure-treated wood are weather resistant. To show the natural grain of the wood, they may be coated with a clear wood preservative.

Decks and porches may also be painted, first by a primer-sealer, then by two coats of porch and deck alkyd or enamel paint. Glossy paints can be slippery when wet. To prevent slipping, a handful of marine sand (very fine sand) may be mixed with each gallon of paint just before application.

Masonry   Masonry such as brick veneer, stone veneer, or flagstone should not be painted because this defeats their advantage of being maintenance free. However, some house styles such as Greek revival or French provincial may be veneered with brick painted white.

Latex paints are preferable since they adhere well to masonry. They are easy to apply and durable. The masonry must be dry before painting.

Steel   Steel must be painted because it is prone to rust. Before painting, the surface should be sanded to remove all rust. Next, a primer capable of neutralizing oxidation (which causes rust) is applied. Finally, one or two coats of finish paints are applied.

Interior Walls and Ceilings   Interior walls and ceilings are usually painted. First, a coat of primer-sealer is applied, preferably before the interior trim is installed in order to avoid smearing the raw wood. Central heating must be in operation if it is done in the winter. Next, one or two coats of latex-, alkyd-, or oil-base paints are applied. If you apply two coats of alkyd- or oil-base paints, walls and ceilings may not have to be repainted for ten years or more.

Interior Doors, Windows, and Trim   It is preferable to stain these items, but it is a lot of work. For one thing, the wood has to be protected when painting the walls and ceilings. Also, the adjacent walls should be protected while staining and varnishing the door and window casings. Staining and varnishing require a lot of sanding, but they produce a first-class finish.

Much easier and cheaper, of course, is to paint all the doors, windows and trim using the same color paints as those of the adjacent walls.

Contracting for Painting and Staining

Painting and staining are usually contracted materials and labor based on competitive bidding. The ratio between labor and materials varies widely. For instance, the labor cost of painting one square foot of drywall stays practically the same regardless of the type and price of the paint chosen for the job.

One gallon of paint covers 300 to 400 square feet of drywall or plaster. A painter can paint about 1400 square feet per day if he uses a roller and double this amount if he uses a spraying machine.

Painting and staining is something you can do yourself, in part or in full, if you have the interest, time, and energy.

CHAPTER 37

Finish Flooring

There are four principal types of finish flooring: (1) wood, (2) hard tiles, (3) resilient tiles, and (4) carpeting. Choosing the type of finish flooring for each area should be based on suitability, initial cost, durability, maintenance, comfort, and appearance.

Wood Flooring

Wood flooring is suitable for bedrooms, family rooms, and living rooms. It may also be installed in hallways. Many people prefer wood flooring because it feels warm underfoot. This is because wood is a good thermal insulator.

What makes wood uniquely beautiful is its natural grain. The beauty of wood can be enhanced by high-quality finishing. Usually, each wood flooring manufacturer recommends a certain type of finish that best preserves, maintains, and enhances the quality of its products.

Wood has two classifications: hardwood and softwood.

Hardwoods Hardwoods are cut from **deciduous trees** that lose their leaves in the fall. Most, but not all, hardwoods are harder than softwoods. Some of the well known hardwoods are: **red oak, white oak, maple, birch, walnut, cherry,** and **mahogany.** The oaks are the most popular, and perhaps the least expensive of all hardwood floorings.

Those who do not worry about the cost can install flooring made of exotic woods such as **merbau, angelique, iroko,** and **mutenye.** Exotic woods can cost up to five times more than oak.

The grades of oak starting with the best are: clear, select, and No. 1 common. The grades of other hardwoods starting with the best are: first grade, second grade, and third grade.

Softwoods Softwoods are cut from **conifers (evergreens).** Some of the popular softwoods are **southern yellow pine, Douglas fir, Ponderosa pine, redwood,** and **western hemlock.** The grades of flooring softwoods are similar to those of rough lumber.

Softwoods are relatively inexpensive. They may be used in the bedrooms of low cost houses; as underlayment for wall to wall carpeting and resilient flooring, in the attic, and inside the closets for moderately-priced houses.

Patterns of Wood Flooring Wood flooring is manufactured in several patterns: **strips, planks, parquet,** and **blocks.**

Strips. Strips are $\frac{25}{32}''$ thick and $2\frac{1}{4}''$ wide, and are sold in random lengths varying from 2 to 16 feet. Most strips have tongue-and-groove edges and grooved bottom (Figure 37.1a). The bottom grooves make the strips more flexible,

reduce the effect of subflooring imperfections on strip leveling, and lessen the transfer of humidity from the subflooring to the strips. Other types of strips include square-edged, and side-and-edge matched.

Planks. Planks are wide strips. They are $\frac{3}{4}''$ thick and 3" to 6" wide. Some planks have $\frac{3}{4}''$ diameter pegs located at each end, which gives them a colonial look.

Parquet. Parquet consists of small strips called **slats** that vary in size and thickness. They can be installed in different patterns (Figure 37.1b) according to the taste of the homeowner.

Blocks. Blocks are parquet slats glued together in the form of blocks that vary in size from 4 × 4 to 9 × 9 inches. The

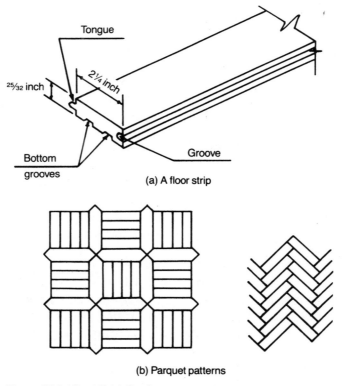

(a) A floor strip

(b) Parquet patterns

Figure 37.1 Wood finish flooring.

edges of the blocks are tongue-and-groove.

Shopping for Wood Flooring The time to shop for wood flooring is when the rough framing is completed. Take the design drawings with you so that dealers can give you estimates for different types of wood.

Delivery and Installation of Wood Flooring Wood flooring should not be delivered until all drywall, plastering, and interior trim (with the exception of base molding) are completed and the house is broom cleaned. Make sure that the dealer is storing the wood indoors under dry conditions. Wood should be delivered about one week before installation and only on dry days. During this week, the wood should be spread out in the area where it is to be installed in order to stabilize its moisture content. The temperature of the house should not be lower than 65°F during storage, and for at least 7 days after installation. A space of about $\frac{1}{2}$″ should be left between the edges of the subflooring and the walls. This space will be covered with base molding.

When strip flooring is to be laid over wood (boards or plywood) subflooring, a layer of building paper (15-pound asphalt-saturated felt) should be laid first on the subflooring. It prevents moisture from getting under the wood strips and also eliminates the squeaking caused by the wood strips rubbing against the subflooring.

When wood strips are to be installed over concrete slabs, the first step is to apply a coat of waterproofing material to the top of the slab. Next, 1 × 4-inch pressure-treated furring strips are fastened to the concrete slab at 16 inches on center spacing (Figure 37.2). The furring strips are then covered with a vapor barrier if one is not installed below the concrete slab. A second course of 1 × 4-inch furring strips is installed on top of the first one. Finally, the strip flooring is installed perpendicular to the direction of the furring strips.

When installing parquet slats or blocks over concrete slabs, all the steps for strip flooring should be followed. In addition, a layer of underlayment and a layer of building paper are installed over the second course of furring strips to provide a base for laying the parquet.

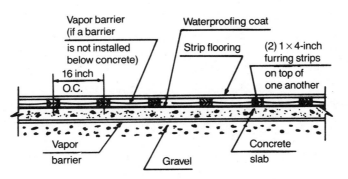

Figure 37.2 Installation of strip flooring over a concrete slab.

Finishing Wood Flooring Wood flooring is sold either prefinished or unfinished. Prefinished flooring must be installed with extreme care to ensure that it does not get scratched. Unfinished wood should be sanded several times in order to make the surface of the flooring very smooth. Sanding starts with coarse sand paper followed by finer paper until the desired smoothness is attained. The flooring must be thoroughly vacuumed after each sanding. Finally, two or three protective coats of varnish or similar material are applied. Sanding and varnishing should be in accordance with the instructions of the flooring manufacturer.

Hard Tiles

Hard tiles include **ceramic tiles, mosaic tiles, quarry tiles,** and **marble.** Homeowners love hard tiles because they are durable, waterproof, easy to clean, and above all, beautiful. When nicely set, they add to the attractiveness and value of the house. Because of their resistance to moisture, building codes require that ceramic or mosaic tiles be installed on the floors of bathrooms and on the walls around the bathtubs and shower heads. In addition to bathrooms, hard tiles may be installed in the kitchen, halls, foyers, and laundry room (mudroom).

Hard tiles are manufactured in two classifications: floor and wall. They look exactly alike except that wall tiles are thinner. Each type of tile has a matching $3\frac{1}{4}''$-wide base molding of the same length and color as the floor tile.

Ceramic Tiles Ceramic tiles are available in a wide variety. There are many manufacturers in this field. Each one has several lines of unique designs and colors. The most common size, and generally the cheapest, is the 4-inch square. Bigger sizes such as 6×4, 6×6, and 8×8 inches are becoming popular even though they are more expensive. Tiles can be bought in different colors; white is the cheapest. Several manufacturers offer lines of solid colors with matching sets of hand painted design tiles. These sets are expensive, and usually only a few are dispersed among the solid-color wall tiles.

Mosaic Tiles Mosaic tiles are manufactured in small sizes: 1×1, 1×2, and 2×2 inches. They are sold with their face glued to 1×1- or 1×2-foot sheets of paper for easy installation. After installation, the paper face is washed away.

Quarry Tiles Quarry tiles are made of clay in solid brown or red colors. They look rugged and are used mainly in kitchens. Their most common size is 4×8 inches, but other sizes are available. Most quarry tiles are $\frac{1}{2}''$ thick.

Marble Tiles Marble tiles vary in price between expensive and very expensive. The least costly is perhaps the **Carara,** which is gray in color. It is imported from Italy. Its price is about double the price of high-quality ceramic tiles. Other marble colors include black, brown, red, and green; these are very expensive. Because of the high price, marble tiles are installed in prominent areas such as the foyer. It should be mentioned that marble is a soft stone, and scratches easily; it must be protected, particularly from dogs.

Marble Saddles Marble saddles, also called **threshold,** are 2-, 3-, 4-, 5- or 6-inch wide marble strips. They are used as sills at door openings where adjacent rooms have different types of finish flooring. For first-class work, the width of the saddles should be equal to the width of the walls at which they are installed.

Shopping for Hard Tiles The appropriate time to shop for hard tiles is when rough framing is almost completed. Hard tile showrooms are packed with a dazzling array of tiles in many sizes, shapes, textures, designs, and price ranges. Take the design drawings of the house with you. Tile dealers can give you valuable suggestions and tips. If they see that you are buying a considerable amount of their product, they will very likely give you a good discount.

Hard Tile Installation Hard tiles should be installed on a rigid foundation, otherwise, they will loosen or break. To install hard tiles over wood subflooring, a layer of black paper should be installed first. Next, $1\frac{1}{2}$"-thick concrete reinforced with one or two meshes of expanded metal lath is poured over the black paper. After the concrete hardens, the tiles and the base molding are installed using a ready-mix cementing compound called **mud.** Finally, the tiles are **grouted** and cleaned.

In lower quality work, the tiles are glued to the plywood subflooring. In this installation, the plywood must be of the exterior type and at least $\frac{3}{4}$" thick. The spacing of the joists supporting the floor should not exceed 16". Installation starts with waterproofing the plywood. Next, the tiles are glued to the subflooring in accordance with the manufacturer's recommendations. The tiles should not be grouted until the volatile substance in the glue evaporates.

Resilient Flooring

Resilient flooring is so called because it is flexible. It is popular because it is inexpensive and easy to install. Many people prefer it for the kitchen because it feels soft underfoot.

Resilient flooring can be made of **vinyl, linoleum, asphalt,** or other materials. It is manufactured in sheets and tiles. The sheets are sold in 6-, 9-, or 12-foot wide rolls, and the tiles in sizes varying from 9×9 to 12×12 inches. Both sheets and tiles are available in a wide variety of patterns, colors, and textures. Many designs are imitations of ceramic tile or wood flooring. Some kinds require waxing, others have glossy (no-wax) surfaces.

Generally, resilient flooring has several disadvantages:

1. It is highly vulnerable to cigarette burns.

2. High heel shoes may damage it.

3. Furniture and appliances cause permanent indentation.

4. Sunlight may cause colors to fade.

5. It is prone to damage from asphalt, which may be tracked into the room from the driveway.

6. Some types, like asphalt tiles, cannot resist grease, and therefore should not be installed in the kitchen.

Manufacturers offer up to a 10-year "limited warranty" for their products. Read the warranty carefully to know exactly what the word "limited" means. For example, some manufacturers exclude high-heel traffic from their warranty!

Shopping for resilient flooring should start after the rough framing is completed. There are many varieties and brand names to choose from. The biggest factor that determines the price of a particular type of resilient flooring is its thickness.

Installation of Resilient Flooring
Resilient flooring may be installed over wood underlayment consisting of $\frac{1}{2}$"- to $\frac{5}{8}$"-thick plywood or particlewood panels, or over concrete slab.

Installation over wood underlayment starts with covering the underlayment with a coat of waterproof adhesive followed by a layer of building paper. Next, the resilient flooring is glued to the building paper. It is important that the flooring seams (joints) do not coincide with the joints of the underlayment. The type of glue and method of installation shall be in accordance with the recommendations of the flooring manufacturers.

The concrete slab upon which resilient flooring is to be installed must have a smooth, level surface! The concrete must be covered with a vapor barrier if one was not laid below the concrete. Furthermore, the concrete must be completely dry before flooring installation begins.

Resilient sheets are preferable to tiles because they have fewer seams. Tiles, on the other hand, are easier to install, particularly those with glue-backing.

Carpeting

Wall-to-wall carpeting is very popular because it is colorful, sound absorbent, impact resistant, and soft underfoot.

Its **plushes,** or **piles,** create a sense of luxury and warmth like no other type of floor covering. Carpeting may be made of **polyester, nylon, acrylic,** or **wool.** Carpets are suitable for bedrooms, family room, living room, and hallways.

Most carpets are sold in 12-foot wide rolls. The length and density of the pile are major factors in determining the price of carpets. The price is also affected by the material of the pile; wool is the most expensive. Carpets are available both in solid colors and design patterns.

Shopping for Carpets Shopping for carpets should start after the rough framing is completed. Visit several showrooms in order to get a good idea of the materials and prices. Take the design drawings of the house with you to show the dealer the amount of carpeting you are buying. This gives her an incentive to offer you a good price.

Carpets are sold per square yard. Dealers give two prices: one for carpet only, and the other for **carpet, padding** and **installation,** all together. Padding may be made of **rubber** or **felt.** Its thickness varies from $\frac{1}{4}$" to $\frac{3}{4}$". Thick padding adds a dimension of luxury to the carpeting: It makes people who walk on the carpet feel that they are sinking into thick, *expensive* carpeting.

Once you agree on the carpets and the prices, you will sign a contract and give a small deposit to the dealer. Subsequently, the dealer sends a representative to your house to take exact measurements. She makes a layout of each room indicating the location of the seams, usually away from the traffic areas. After you approve the layout, the dealer calculates the exact price, and asks for a more substantial deposit. In a few weeks, the carpet is cut in the shop, and ready for delivery.

Installation of Carpeting   Installation begins with laying **tackless wood strips** around the perimeter of the room. Next, the padding is laid within the strips and stapled to the floor every few inches in both directions. Finally the carpets are laid, stretched, and fastened to the tackless strips. The pieces are joined together by means of an adhesive tape laid under the seams. An electric iron is used to heat the tape causing the adhesive to melt. Next, the edges of the carpeting are pressed against the tape for a few minutes until the adhesive dries. When professionally done, the seams can hardly be detected.

Contracting for Finish Flooring

Finish flooring may be contracted material and labor, or you may buy the material, and contract labor only. Generally, it is a good idea to contract material and labor on fixed-price basis. But if you think that you might change your mind about the materials, it is best to buy the materials and contract labor only.

CHAPTER 38

Bathrooms and Kitchen

The bathrooms and kitchen are important features of any house. In addition to their functions, they contain the house's largest decorative items and appliances: bathtubs, vanities, lavatories, toilets, skylights, ceramic tiles, mirrors, kitchen cabinets, refrigerator, oven, dishwasher, garbage compactor, and so forth.

There is a growing tendency among prospective homeowners, particularly those who contemplate big houses, to spend a great deal of money on their bathrooms and kitchens. They want them bigger and more luxurious. This gives them something tangible to show off and enjoy. Spending money on bathrooms and kitchens is also a sound investment. Each dollar spent on bathrooms and kitchens increases the value of the house by more than one dollar.

Bathrooms and Powder Rooms

A bathroom should include a **toilet**, a **bathtub**, a **lavatory**, a **vanity**, and a **mirror**. A **powder room** is designed for guest use and does not include a bathtub.

Only one bathroom is required to obtain the certificate of occupancy. For practicality, however, almost every house includes more than one bathroom. Small houses have 1½ bathrooms (a half bathroom does not include a bathtub); medium-size houses have 2 or 2½; and big houses have 3 or more bathrooms.

Bathroom Floor and Wall Finishes

As mentioned in Chapter 37, the floors of the bathrooms and powder room and the walls surrounding the bathtubs and shower heads must be finished with ceramic or mosaic tiles extending a minimum of six inches above the shower head. Wall tiles should be installed over wonderboards instead of drywall.

In many localities, you are required to install a marble saddle at the bathroom entrance. The saddle should be about one inch higher than the bathroom floor; its purpose is to prevent water from spilling over into the adjacent area in case of flooding.

Bathtubs Bathtubs are the biggest feature of bathrooms. They are sold in different shapes, sizes, colors, and materials.

The standard bathtub is rectangular in shape. It is 54 to 60 inches long, 30 to 32 inches wide, and 16 to 18 inches deep. Most are made of cast iron with baked enamel finish. Low cost bathtubs are made of fiberglass with acrylic gloss finish. The least expensive color (for all plumbing fixtures) is white. The shower head is usually installed over the bathtub.

Luxury bathtubs are available in rectangular as well as oval shapes of up to 84 inches long. Some of them are provided with grab bars for safety and comfort. Others have built-in seats and ledge spaces. A shower head should not be installed over a luxury bathtub, but rather in a separate shower enclosure.

The elegance of a bathtub can be enhanced by installing it sunken within a raised platform that is accessed by wide steps. The platform and steps are finished with ceramic or mosaic tiles matching those of the floor and wall of the bathroom. (Marble is not recommended since it is not water resistant.)

Whirlpools and Spas Most luxury bathtubs include a whirlpool as an option. (Some people call it a **Jacuzzi**, which is a tradename). A whirlpool consists of several water jets operated by a pump.

Spas are deep bathtubs with whirlpools. They are mostly square, but they come in other shapes as well.

Both whirlpools and spas may be made of cast iron with baked enamel finish or reinforced fiberglass with gloss acrylic finish. The cast iron units are durable and their colors do not fade, but they are heavy and expensive. Fiberglass units are much lighter and less costly, but their colors may fade, and some types cannot stand hot water.

Lavatories, Toilets, and Bidets Lavatories, toilets, and bidets are sold in many different shapes, materials, and colors. Visit several showrooms before choosing any plumbing fixtures.

Lavatories are usually installed over vanities. Many building material suppliers and big department stores offer economy units consisting of a lavatory and a vanity molded together. On the other end of the scale there are luxurious free-standing pedestal lavatories. They are very elegant, and are usually reserved for powder rooms.

A **bidet** is a luxury item that is installed only in the more expensive houses.

Vanities, Mirrors, and Medicine Cabinets Vanities not only accommodate the lavatories, but they are also used to hold shaving articles, makeup, and other toiletries. They have drawers and cabinets for storage of larger items. Vanities are usually made to order by a kitchen cabinet dealer or a remodelling carpenter; most are made of particleboard veneered with Formica. They are delivered to the house complete and ready for installation, after which the carpenter cuts the top out to accommodate the lavatory.

Each vanity must have a mirror

either by itself or combined with the medicine cabinet.

Skylights   Installation of skylights in bathrooms is a feature of contemporary style houses, and the idea is spreading to other styles.

Kitchens

Kitchens vary considerably in size, depending on the size of the house. For a small house, a kitchen may be 9 feet wide by 11 feet long. Some homeowners prefer to have the kitchen open to the family room with a service counter in between. This arrangement makes both the kitchen and the family room look bigger.

For large houses, the size of the kitchen can be 12 × 25 feet or even bigger. A large kitchen may include a **dinette,** also called a **breakfast nook,** so that the dining room may be left for formal occasions.

Kitchen Cabinets   Kitchen cabinets are a main feature of any kitchen. In addition to storing the dishes, cookware, and silverware, they are a major decorative item.

Kitchen cabinets may be either prefabricated (**stock**), or custom made. Stock cabinets are sold by cabinet dealers, building materials suppliers, and major department stores. They exhibit quite an array of cabinet styles and designs. Custom made cabinets can be built by finish carpenters either in their shops or on site, according to the exact dimensions of the kitchen.

Almost all kitchen cabinets, whether stock or custom made, are fabricated of wood, wood products, and Formica. The price of the cabinets depends on their material, construction, detailing, and finishing. Popularly priced cabinets are made of particlewood veneered with a thin layer of hardwood or Formica. More expensive cabinets have solid hardwood doors and drawer fronts, and hardboard sides, backs, and shelves. Some hardwoods are more expensive than others. Oak is perhaps the least expensive; cherry is one of the most expensive woods. The better cabinets are finished with stain and lacquer, as is any high quality furniture.

Contemporary style kitchen cabinets are becoming popular. The front is made of plywood with solid-color Formica veneer. The front of the refrigerator and other kitchen appliances may be covered with matching veneer.

Kitchen cabinet accessories include a **lazy Susan** corner cabinet with two or three circular metal trays. The door and trays revolve around a vertical axis to provide easy access to the items shelved deep inside the cabinet. There is also the **pantry** unit, which consists of several sets of folding shelves. It provides a considerable amount of storage area without occupying too much space.

Stock cabinets are sold in standard sizes. **Base cabinets,** which are installed on the floor, are 36 inches high by 24 inches deep (Figure 38.1). **Wall cabinets,** which are attached to the walls, are 12 to 30 inches high by 12 inches deep. The width of base and wall cabinets starts from 10″ and increases by 2″ increments until it reaches 48″. Small gaps between adjacent cabinets are closed by **filler strips.**

Installation of Kitchen Cabinets   Kitchen cabinets are installed after the drywall is taped, or the plastering is dry, and before the finish flooring is installed. The base cabinets should be laid perfectly horizontal leaving a vertical gap between them and the subflooring. The size of this gap is equal to the thickness of the finish flooring. For ceramic tiles, it is about two inches, which is the thickness of the concrete base and the tiles. For resilient tiles, the gap is a fraction of an inch.

After the base cabinets are installed and leveled, the dealer's representative takes the exact measurements of the cabinet tops in order to fabricate the **counter tops.** For good quality work, all adjacent cabinets should be covered with one counter top. Counter tops are usually made of particlewood or plywood veneered with Formica.

In the meantime, the carpenter installs the wall cabinets and the **back splash,** which is the molding laid against the wall over the base cabinets.

There are two types of back splash: **half back splash,** which is about 4" high, and **full back splash,** which covers the entire height between the base and wall cabinets. The latter is recommended for sanitary reasons.

Finally, the carpenter makes cutouts in the counter top to accommodate the range top and the sink, and in the full back splash to accommodate the receptacle outlets.

Kitchen Appliances The **oven** and **range** is the only appliance that is required to obtain the certificate of occupancy. Some ovens have self-cleaning or continuous-cleaning features, and timing clocks. A large kitchen may have a **range,** a conventional **oven,** and a **microwave oven.**

Second in importance is the **refrigerator.** The least expensive units consist of a cooling compartment and a

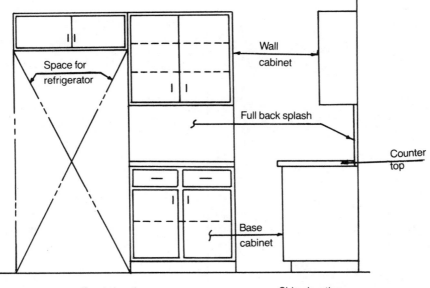

Figure 38.1 Kitchen cabinets.

freezer. Modern refrigerators are designed with two compartments side-by-side, one for cooling and the other for freezing. The freezer side usually includes ice and cool water dispensers. Some refrigerators even make ice cream!

Dishwashers have become very popular. Modern units have electronic dashboards for selecting the length of the washing cycle and the water temperature.

A **garbage compactor** is usually found in the more expensive houses.

Contracting for Kitchen Cabinets

Kitchen cabinets should be contracted as soon as the drywall or blue boards are installed. The design drawings will probably show the arrangement of the kitchen cabinets, but this should be used as a starting point only. The final arrangement of the kitchen and the style, color, and construction of the cabinets should be decided on after you and your family visit several showrooms, and discuss your requirements with the cabinet dealers. The dealer usually sends a representative to your house to take the exact measurements of the kitchen with all the features that may affect the arrangement of the cabinets. Based on this information, the dealer creates two or three designs indicating the unit sizes, styles, and the type of wood, etc. She then gives you a price which should include the counter tops and back splash. The cabinets may be installed by either the trim carpenter or the dealer's carpenter.

You should make your buying commitment early, since kitchen cabinets may take four to eight weeks for delivery. As soon as the dealer notifies you that your cabinets are ready, check with the trim carpenter (if he has the installation contract) to see that he is ready to install them as soon as they are delivered. Kitchen cabinets should not be allowed to sit around since they are an easy and tempting target for theft.

CHAPTER 39

Sewage Disposal and Water Wells

*I*f your area is not served by a public sewer line, the household sewage has to be disposed of by a local sewage disposal system consisting of a **septic tank,** a **distribution box,** an **absorption field,** or a **seepage pit** (Figure 39.1).

The Separation Distances

To minimize health hazards and water pollution, the health department of each state establishes minimum **separation distances** between the edge of the house and the nearest edge of the septic tank; between the edge of the house and the distribution box; and between the absorption field and the nearest water well or water pond. These distances depend on many factors including the type of soil, the topography of the area, and the type of disposal systems (absorption fields or sewage pits). As an example, the separation distances shown in Figure 39.1 are

those specified by the department of health of the state of New York.*

Septic Tank

A septic tank is a rectangular chamber in which the sludge of the household raw sewage decomposes and settles. The capacity of the septic tank is determined by the number of bedrooms of the house, or the expected number of its occupants, and by the regulations of the department of health of each state. In no case should the capacity of the septic tank be less than 500 gallons.

Septic Tank Specifications The length of the tank should be between two and four times its width. The depth of sewage should not be less than four nor more than five feet. The **invert**

Waste Treatment Handbook, Individual Household Systems, New York State Department of Health.

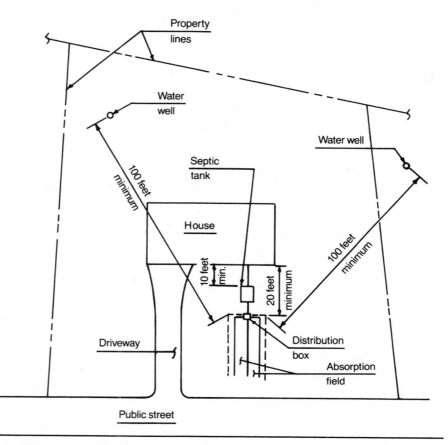

Figure 39.1 Absorption field separation requirements.

(lowest point) of the outlet pipe must be 2″ below the invert of the inlet pipe (Figure 39.2) to prevent sewage from backing up into the house. Uniform flow of sewage across the width of the septic tank is effected by two **baffles**: one after the inlet pipe, and the other before the outlet pipe.

Septic Tank Construction   The septic tank may be constructed of poured concrete or prefabricated fiberglass. The latter must be delivered to the site fully assembled. Whether made of concrete or fiberglass, the septic tank should be installed over a four-inch thick layer of gravel or crushed stones. The tank top should be covered with about 12″ of soil. Its roof should include two covered openings: one for inspecting the tank and the other for cleanout purposes.

Distribution Box

The distribution box distributes the effluent of the septic tank uniformly to the trenches of the absorption field.

Distribution Box Specifications   When looking at it from above, the shape of the distribution box may be

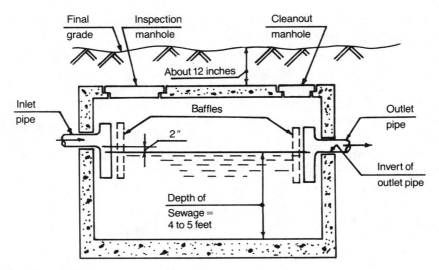

Figure 39.2 Septic tank.

circular or rectangular. To ensure uniform distribution of effluent, the invert elevation of all outlet pipes must be exactly the same. To prevent the effluent from backing up into the septic tank, the elevation of the invert of the outlet pipes must be 2" below the invert of the inlet pipe.

Distribution Box Construction The distribution box may be constructed of concrete or fiberglass. The box should be constructed over a 12-inch thick layer of clean sand or pea gravel ($\frac{3}{8}$" maximum size). The top of the distribution box must be covered with 12 inches of soil. Its roof must have a removable cover for inspection.

Absorption Field

An absorption field consists of several absorption trenches. The effluent is transmitted to the trenches through pipes made of perforated plastic or clay laid end to end. The total length of the absorption trenches is determined by the daily amount of effluent and the results of the percolation test (see Chapter 17). The number and direction of the absorption field trenches is determined by the size of the lot and the slope of the ground.

Absorption Trench Specifications The absorption trenches are 24 inches wide and 18 to 24 inches deep. Their bottom should be slightly sloped ($\frac{1}{32}$" to $\frac{1}{16}$" per foot). There should be a minimum vertical distance of two feet between the bottom of the trench and high seasonal ground water, bedrock, or impervious soil (such as stiff clay).

Absorption Trench Construction The trenches are excavated and filled with a six-inch thick layer of gravel or crushed stones. Next, the drainage pipes are laid in the middle of the trenches and covered with gravel or crushed stones to a height of two inches above their top. At this stage, the absorption field may have to be inspected by the local health department. After inspection, a layer of building paper is laid over the gravel and the trenches are backfilled. The purpose of the building paper is to prevent the backfill from clogging the pores of the gravel or crushed stones.

Heavy vehicles and equipment must not be allowed over the absorption field since they will compact the gravel in the trenches and may cause the drainage pipes to misalign or break.

Seepage Pit

A seepage pit, also called **leaching pit,** may be used to discharge the effluent of the septic tank if the construction of an absorption field is not feasible.

Seepage Pit Specifications A seepage pit consists of a circular hole with a roof. The walls of the hole may be lined with open-jointed concrete blocks, precast concrete rings, or stones. The size of the pit (or pits) is determined by the daily amount of effluent and the results of the percolation test (see Chapter 17). The effluent is discharged to the surrounding soil through the joints of the wall lining and the open bottom of the pit.

Seepage Pit Construction The construction of the seepage pit starts with excavation. The diameter of the excavation should be at least one foot larger than the outer diameter of the wall lining. The bottom of the pit should be at least two feet higher than the highest level of underground water, bedrock, or impervious soil (check with the department of health of your state in this

regard). Next, the wall lining is constructed. The space between the excavation and the lining should be filled with gravel or crushed stones, the purpose of which is to laterally support the lining and facilitate the discharge of the effluent. The top of the pit should be covered with 12 inches of soil. Its roof should have a removable cover made of precast concrete.

Water Wells

Water wells provide the household water in areas that do not have public water systems. There are two types of water wells: shallow wells, and artesian wells.

Shallow Wells A shallow well, also called **gravity well,** draws water from close to the surface aquifer. An **aquifer** is a water bearing stratum of permeable soil such as sand, gravel, or rocks. The depth of a shallow well varies from 15 to 35 feet. The water is pumped out by a pump located on top of the well or in the basement. It is to be noted that surface pumps draw water by suction, and thus cannot draw water deeper than about 33 feet which is the atmospheric pressure. Shallow wells are not permitted where household sewage is disposed of through seepage pits (check with the department of health of your state in this regard).

Artesian Wells An artesian well, also called **pressure well,** draws water from a confined aquifer containing water under pressure. The water pressure is produced by a high water level at a distant hill or mountain (Figure 39.3). The water pressure may be high enough

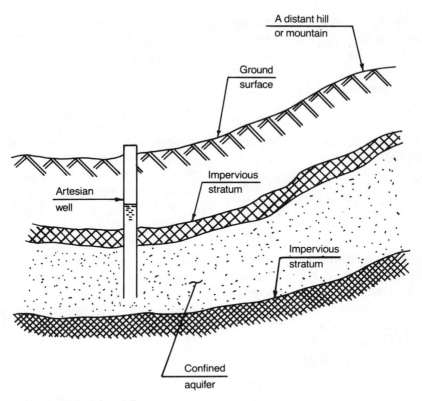

Figure 39.3 Artesian well.

to cause water to flow freely off the well's head. However, with most wells the water must be pumped out.

Drilling the Well Well drilling requires heavy equipment since most wells are 50 to 200 feet deep. A hole is drilled deep into the ground and a steel **casing** is lowered into it. The function of the casing is to prevent the surrounding soil from caving into the hole. The hole should penetrate into the water bearing stratum until it produces a sustained yield of about 5 gallons per minute.

The depth of the casing depends on the type of the penetrated stratum, the regulations of the local department of health, if any, and the common prac-

tice in the area. The top of the casing should extend above the final grade elevation by at least 12 inches. The ring of space between the hole and the casing must be filled with cement grout to prevent pollution from flowing into the well. After drilling is completed, the water pump is installed and the top of the well is sealed.

Well Log The well contractor should provide you with a well log, which is a record of the depth and diameter of the well, the casing depth and diameter, the depth of the cement grout, the well yield, the types and depths of the formations penetrated, and the depth of the water bearing stratum.

Pressure Storage Tank A pressure storage tank is required to meet the water peak demand. The tank capacity is a function of the size of the house and the sustained yield of the well.

Contracting for Sewage Disposal and Water Well

Constructing sewage disposal systems and water wells are two distinct professions.

Sewage Disposal System A sewage disposal system should be contracted on fixed-price basis. Solicit bids from three reputable contractors. Each bid should include the size and material of the septic tank, distribution box, absorption field, or seepage pit. The design and construction of all of these must conform to the regulations of the department of health of your state.

Water Well It is recommended that the water well be contracted on fixed-price basis. Solicit bids from three reputable contractors. Each bid should include the minimum yield of the well, the minimum depth of the casing, the depth of cement grouting, the type and capacity of the water pump, and how the well will be disinfected.

Because the exact depth of a well cannot be determined in advance, contractors usually give a price for a depth up to a certain amount, say 200 feet, and a price per linear foot thereafter. This is to cover the costs of moving the heavy equipment to and from the site.

CHAPTER 40

Decks, Porches, and Patios

Decks, **porches,** and **patios** greatly enhance the pleasure of owning a house. When furnished with chairs, a table, an umbrella, and a grill, this outdoor space allows you and your family, friends, and relatives to enjoy cooking, dining, and relaxing in the privacy of your own back yard. In itself, being able to sit outdoors in nice weather alleviates the effects of "cabin fever" brought on by staying indoors during the winter season.

The location, size, and shape of the deck, porch, or patio are determined by the size and design of the house, the size and shape of the lot, your budget, your preference, and the local zoning ordinances. Usually, the rear and side yards (see Chapter 17) are measured from the property lines to the outer edges of the decks and porches.

Decks

A deck is a raised platform. It may be accessed from the family room, kitchen, or both.

Deck Construction Most decks are of post and beam construction. The outer edge of the deck is supported by piers and posts (Figure 40.1) spaced at 12 feet, maximum. The piers and their footings may be constructed of ordinary concrete, bricks, or stones. After the piers harden, the posts are installed. They may be made of wood, steel pipes, concrete, brick, or stone. A built-up girder consisting of three two-by-tens is installed over the posts. The inner edge of the deck may be supported by a ledger fastened to the wall of the house (Figure 40.1), or by piers and posts. The bottom of the pier foundation must be at the same level as that of the house. The next step is to install 2 × 10-inch joists spaced at 16 inches o.c. over the built-up girder and ledger. Finally, decking consisting of 2 × 6-inch planks is laid over the joists. There must be a space of $\frac{1}{4}''$ between adjacent planks to allow rainwater and melting snow to pass through.

For safety, decks should have balusters and railings along their periphery. Deck railings can be of any design

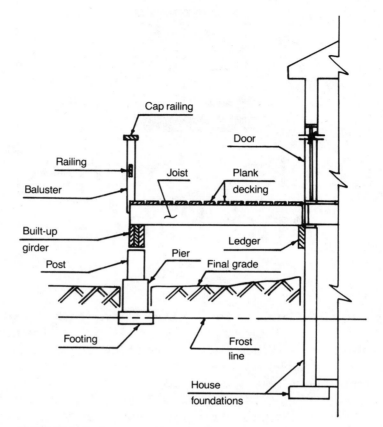

Figure 40.1 Deck components.

you wish. Some homeowners prefer wide cap railings that allow them to place potted plants or electrical fixtures on top.

The deck should be accessed from the ground by one or two stairways, depending on the size of the deck. Deck stairs are usually of the open riser type, having treads but no risers. The width of the treads and the height of risers must follow the specifications outlined in Chapter 32. A concrete pad is poured at the bottom of the stairs. It serves as a landing, and to support the bottom ends of the carriages. All lumber used for deck construction must be pressure-treated to withstand the weather.

Porches

A porch is similar to a deck in that it is a raised platform supported by piers and posts. However, it has a roof, and the front and sides may be partially or totally enclosed.

Porch Construction First, the piers, posts, built-up girders, and joists are constructed in a similar way to those of a deck. Decking is then constructed of tongue-and-groove or shiplap boards laid with no space in between. Next, several columns are installed over the outer edge of the decking to support the outer edge of the roof. The roof framing and covering should match, or at least be compatible with, those of the house.

Patios

Unlike decks and porches, **patios** are constructed on the ground. Usually, the rear and side yard zoning ordinances do not apply to patios.

Patio Construction   Patios consist of a surface and a base. The surface may be constructed of ordinary or reinforced concrete, flagstone, or bricks. The construction of the base depends on the type of the surface. For concrete surface, the base consists of four-inch thick gravel.

For flagstone or brick surface, the base consists of a four-inch thick concrete slab laid over four inches of gravel. It is a good idea to reinforce the concrete base in cold areas. If the soil is sandy or rocky and will not settle, the base may consist of four inches of gravel only.

Contracting for a Deck, Porch, and Patio

Deck and porch contracting involves a mason, a carpenter, and perhaps a roofer. The piers and footings are contracted to a mason, usually on a fixed-price basis. The wood framing is contracted to a carpenter either labor and material or labor only. As with the wood framing of the house, labor should be about equal to the price of the lumber. The roofing may be contracted to a carpenter or a roofer per square (100 square feet of exposed surface), or on a fixed-price basis.

A patio is contracted to a mason either labor and material or labor only. The contract should specify the type and thickness of the base and the surface. Contracting may be on fixed-price basis or per square foot.

CHAPTER 41

Site Grading, Lawn, and Landscaping

Site grading, lawn, and landscaping begin after the house is completed. The site is transformed from piles of dirt into a beautiful green carpet of lawn and an eye-pleasing arrangement of shrubs and trees.

Site Grading

The purpose of site grading is to make the site as smooth as possible and to prepare it for driveway and walkway construction; for lawn seeding or sodding; and for landscape planting.

The machines usually used for site grading are the bulldozer, backhoe, and front end loader (see Chapter 19). Ideally, the site should be graded without the need for importing or trucking away soil. The soil resulting from the excavation of the foundations and basement should be used to raise the grade elevation around the house in order to drain away surface water.

Lawn

Lawn is the green carpet of grass that covers the lot. In addition to beautifying the lot, lawn prevents soil wash, refreshes and cools the air in summer, and provides you and your pets with a soft, clean surface to walk and play on. Lawn is planted after the final grading is completed and the driveways and walkways are constructed.

Lawn Seeding and Sodding Lawn may start from seeding or sodding. Seeding is the sowing of grass seeds. Sodding is the laying of ready-bought mats of grass.

The best time to sow the seeds is late summer and early fall. To prepare the soil for seeding: (1) boulders are removed and replaced with loose soil; soil lumps are broken loose, (2) a four- to six-inch thick layer of top soil mixed with peat moss is spread and raked, (3) lime and fertilizers are applied with a spreader and raked, (4) the soil is watered for a week, then raked, and (5) the seeds are sown using a spreader and raked.

The amount of lime depends on the acidity of the soil; and the type, amount, and application of the fertilizers should be in accordance with the manufacturer's recommendations. After the seeds are sown, the soil should be lightly watered once or twice a day until the lawn is established. The seeds should be a mixture of different species so that if disease strikes one, the others survive.

Preparing the soil for sodding is similar to the process for seeding except that the thickness of the top soil may be three to five inches only. Sod is sold mostly in $1\frac{1}{2}$ × 5-foot mats or strips. Its biggest advantage is that it immediately provides the lot with a thick healthy lawn. However sod is expensive. After laying the sod, it should be watered generously the first night, then every evening for 14 days.

Landscaping

Landscaping is the arrangement of shrubs and trees in a tasteful and beautiful design. Most shrubs used for landscaping are evergreens, meaning they do not lose their leaves in the fall. The amount, species, and size of shrubs and trees are determined by your budget, the size of the house, and the size of the lot.

Planning the Landscape If this is your first house, start by visiting the local nurseries to learn the names, characteristics, and prices of different species of shrubs and trees. Next, tour the neighboring area to get an idea about different landscape arrangements. If you like a particular arrangement, count the shrubs and trees and notice their sizes to get an idea of the cost. Based on your observations and budget, you can determine with reasonable accuracy the number and size of shrubs and trees you can afford.

It is a good idea to obtain professional help in landscape planning. For a big job, you may retain a landscape architect or designer. If the job is small or medium in size, you may seek the services of a staff member of one of the larger nurseries.

Taxus, Azaleas, and Rhododendrons Taxus, azaleas, and rhododendrons are the most popular evergreen shrubs. Here is a brief description of their characteristics and uses:

Taxus: Planted mainly around the house and as a fence. When planted at 2'-6" intervals, they grow into a continuous green wall. They are not flower-bearing.

Azaleas: Very popular shrubs that flower in spring. Their flowers may be white, pink, red, or violet. Some species are 18 to 24 inches high and may be planted at 2'-6" intervals to form a wall around the house or in front of a row of taxus. Other species are 3 to 4 feet high, and may be planted at the sides of the main entrance door or anywhere on the lawn.

Rhododendrons: Another popular shrub that flowers in the spring. Their flowers are bigger and last longer than azaleas. Some species require shade while others do better in full sun. Their height varies from 2 to 8 feet. They may be planted near the house or anywhere on the lawn.

Landscaping as an Investment Many people do not realize that landscaping is a good investment. First, it immediately increases the value of the house. Second, when moderately sized shrubs and trees are left to grow, their value increases significantly in a few years. For example, a small rhododendron that costs $30 will be worth twice that amount in four or five years.

The same is true for trees. One of the most beautiful flowering trees is the **kwanzan cherry**. It blossoms early in the spring with large pink double flowers. A nearby nursery is offering small ones for $60, and large ones for $500 each. When I asked the nurseryman about the difference between them, he told me that the only difference is that the big trees are 10 years older than the small ones. Thus, if a kwanzan cherry is bought while it is still small and allowed to grow, it will increase in real value (not allowing for inflation) by 733 percent in 10 years. This is proof that money can grow on trees!

Contracting for Site Grading, Lawn, and Landscaping

Site grading may be assigned to an excavation contractor on a fixed-price basis, or you may rent a machine and operator per day.

Lawn seeding may be contracted to a gardener or a landscaper on a fixed-price basis, or at an hourly or daily rate. Lawn sodding should be contracted sod and installation on a fixed-price basis.

It may be in your interest to buy the shrubs and trees from a nursery and hire a gardener or someone from the nursery to plant them. The cost of planting is about half the price of the shrubs and trees. You should know that most nurseries conduct two sales each year, one in the spring and the other in the fall. A sale would be a good time to buy the bulk of your shrubs and trees.

Landscaping may also be contracted on a lump sum basis. In this case, the landscaper must specify the number, size, and species of all the shrubs and trees included in his price.

CHAPTER 42

Driveways and Walkways

A driveway is the clean and durable surface that a car uses to reach the garage from the public road. Expensive houses have in addition circular driveways that enable drivers to drop their passengers off at the steps of the main entrance door (Figure 42.1). Driveways are constructed after grading the site and before seeding or sodding the lawn.

Driveway Layout

First, the driveway has to be laid out according to the size and shape of your lot and the location of the garage door. The minimum width of a driveway should be 10 feet, but 12 feet is preferable. The inner radius of a driveway turn should be five feet, minimum. To give cars room to get in and out of a garage that has the door on the side, a 25- to 30-foot wide turnaround area is required in front of the garage door (Figure 42.1). This turnaround area should be sloped outward by $\frac{1}{4}''$ per foot to discharge the rainwater away

from the garage. The slope of the rest of the driveway should not exceed 10 percent. If the garage is lower than the surrounding grade, a drain covered with grating should be installed at the entrance of the garage to collect rainwater and melting snow.

You should check with the public works department in your area regarding driveway regulations. Some localities require a minimum distance between the edges of the driveway and the property lines. Another regulation may be that the street curb at the driveway entrances be cut and shaped only by a bonded contractor.

Driveway Construction

The simplest method of constructing a driveway is to lay a six-inch thick course of gravel or crushed stones on the ground. However, this surface is hard on cars and creates dust. Most driveways are constructed of asphalt, and a few are made of concrete.

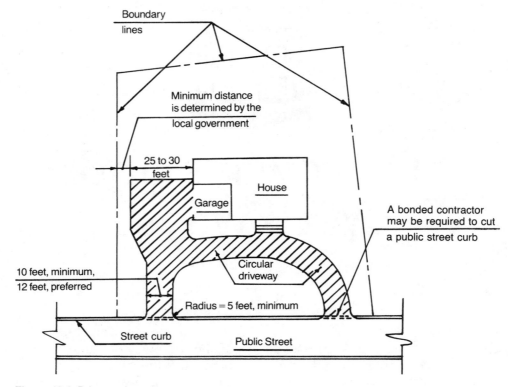

Boundary
lines

Minimum distance
is determined by the
local government

25 to 30
feet

Garage

House

A bonded contractor
may be required to cut
a public street curb

10 feet, minimum,
12 feet, preferred

Circular
driveway

Radius = 5 feet, minimum

Street curb

Public Street

Figure 42.1 Driveway layouts.

Asphalt Driveway An asphalt driveway is called **flexible pavement** because it is plastic in nature and can take some amount of deformation without breaking. It consists of a base course made of gravel or crushed stones, and a surface course made of asphalt mix.

The soil is graded and the gravel or crushed stone base course is laid, no less than three inches thick. The gravel is then compacted by a heavy machine roller to ensure that the pavement will not settle and crack in the future, and to produce a smooth surface for laying the surface course. Just before laying the surface course, a coat of liquid bitumen is applied to the base course to bond the two courses together. The surface course is laid and finished smooth, then sealed. The top of the asphalt course should be crowned in order to discharge the rainwater off the driveway. Asphalt driveways should be sealed every 3 to 4 years.

If the soil is too soft, it is a good idea to drive on the base course for a few months to allow the soil to settle before applying the surface course.

Many homeowners install a curb made of stones or bricks along the periphery of the driveway to prevent the lawn from encroaching on the asphalt.

Concrete Driveway A concrete driveway is called **rigid pavement** because it will crack under very little deformation. It consists of a base course made of gravel or crushed stones, and a surface course made of concrete.

Construction begins with grading the soil. Next, a five-inch thick course of gravel or crushed stones is laid. Finally, a five-inch thick concrete slab is poured. It is recommended that the slab be reinforced with wire mesh to control cracking. To prevent concrete cracking due to temperature changes, expansion joints must be provided every 40 feet, at the edge of the turnaround area, and at the garage entrance. To prevent cracks resulting from concrete drying, construction joints must be installed at 10-foot intervals. All joints must be filled with bituminous material to prevent rainwater from getting under the concrete. Concrete driveways are expensive, but, if properly constructed, can last a long time without maintenance.

Walkways

Walkways are the clean and durable surfaces that people walk on. They may be constructed of concrete, bricks, flagstone, or asphalt.

Layout of Walkways Walkways are laid out to link the entrances of the

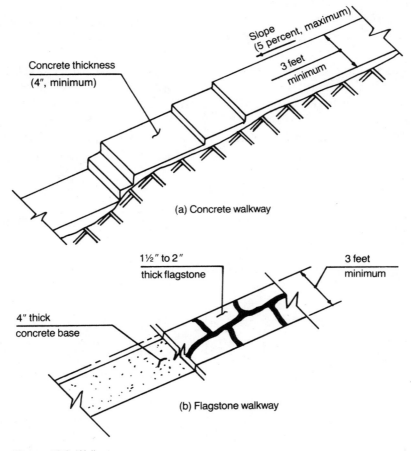

Concrete thickness (4", minimum)

Slope (5 percent, maximum)

3 feet minimum

(a) Concrete walkway

1½" to 2" thick flagstone

3 feet minimum

4" thick concrete base

(b) Flagstone walkway

Figure 42.2 Walkways.

house and the porch, deck, or patio with the public street, or paved driveway. The width of walkways should not be less than three feet, and their slope should not exceed five percent. If the grade is steep, walkways should be stepped as shown in Figure 42.2a. All walkways should be cambered (arched) to prevent water from accumulating on top.

Walkway Construction　Concrete walkways should be no less than four inches thick. They should be poured over a four-inch thick layer of gravel or crushed stones unless the soil is rocky or sandy and is not expected to settle. The concrete should be provided with reinforcement and expansion joints as for concrete driveways except that walkway expansion joints should be spaced at only four feet.

Flagstone or brick walkways should be laid over a four-inch thick concrete base (Figure 42.2b). It is recommended that the concrete be reinforced with wire mesh to resist differential settlement.

Construction of asphalt walkways is similar to that of driveways. They should be included in the asphalt driveway contract.

Contracting for Driveways and Walkways

Driveways, walkways, and the driveway curb are contracted separately. Asphalt driveways are assigned to a paving contractor. Concrete driveways and walkways are contracted to a mason. The driveway curb is usually contracted to a mason. The unit of pricing for driveways and walkways is the square foot, and for the driveway curb is the linear foot.

For each item, invite three reputable contractors, one at a time, to meet you at the site. Explain to each of them what your plans are and ask for suggestions. After you make the final decision regarding the layout, widths, and thicknesses of each item, ask each contractor to give you a fixed price. Contract awarding should be based on the contractor's price and reputation.

CHAPTER 43

Certificate of Occupancy and Tax Assessment

Before you can move into the house, you must obtain a **certificate of occupancy (c. of o.)** from the local building department. You will have to submit to them a copy of the final survey and the fire underwriters (certificate of electrical inspection), and the house must pass a final inspection.

Final Inspection

As soon as the house is ready for occupancy, call the building inspector and ask for an appointment for the **final inspection.** Both the inspector and yourself tour each part of the house: the bedrooms, living room, family room, bathrooms, kitchen, ga-

rage, basement, boiler area, fireplace areas, attic, etc. to check that each item is built in accordance with the building code. He or she specifically checks that all the stairs have handrailings; that the door between the garage and the house is fireproof and is furnished with a self-closing spring; and that the wood framing in the vicinity of the boiler is covered with drywall. The inspector also checks the exterior to ensure that the zoning ordinances are complied with.

Certificate of Occupancy

If the inspection is satisfactory and the house is constructed in accordance with the rules and regulations, you will be mailed a **certificate of occupancy** within a few days.

> *This certifies that (your name), the owner of the premises described hereinafter, has complied with all the provisions of the zoning ordinances and the building code of (city, town, or village), and is hereby permitted to use the structure located on (brief land description) for use as (one family dwelling only) known as (the street address), and that the building is constructed in accordance with the plans filed under permit number (No. of the building permit).*

Temporary Certificate of Occupancy

If the inspector finds violations, they will be brought to your attention. If these violations are minor and do not endanger the safety of the occupants, you will be issued a temporary c. of o. valid for a month, renewable, which allows you to move into the house provided that you correct the violations. When this is done, you invite the inspector back once more to check only the items that needed fixing. After seeing that the violations are corrected, the inspector issues a certificate of occupancy.

Changing Insurance Policy and Tax Assessment

Once you have obtained the c. of o., you must change the insurance policy from builder's risk to homeowner's.

When the value of your property is reassessed, you will get a notice to this effect in the mail. Usually, the tax assessors check on all the properties annually to record new improvements. When he observes that you have put up a house, the tax assessor records this improvement and increases the assessed value, and accordingly, the taxes on the property.

If you think that the taxes levied on your house are too high, you have the right to inspect the tax records of the neighboring houses, and use this as a base for comparison. If you and the tax assessor do not agree, you may file a complaint with the local board of tax review. If you do not agree with the board's decision, you may take your case to court.

The certificate of occupancy is a very important document that must be kept in a safe place. You will need it to get the final installment of the construction loan, a long-term mortgage loan, and possibly when the time comes for you to sell the house.

Glossary

Abstract of title A history of all recorded documents that affect the title to a property, arranged in the order in which they were recorded.

Abstracter (conveyancer) An expert in title search and abstract preparation.

Acceleration clause A clause giving the lender the right to declare that the remaining balance of the loan is due immediately; usually exercised when the borrower defaults.

Acknowledgment A formal declaration before a duly authorized officer (such as a public notary) that a document was signed willingly.

Acre A unit of land measurement equal to 43,560 square feet.

Action of specific performance A court action to compel a defaulting party to comply with the provisions of a contract.

Actual notice Knowledge of facts based on what a person has seen, heard, or read.

Adjustable rate mortgage (ARM) A mortgage loan in which the interest rate is tied to a certain monetary index, and changes upward or downward to follow this index.

Agent A person (such as a real estate agent) authorized by a principal to transact or manage some business on his or her behalf.

American Arbitration Association Establishes the rules for arbitration and provides a list of arbitrators (for a fee).

Amortized loan A mortgage loan that is paid in periodic installments that include interest and part of the principal so that the principal will be paid in full at the end of the term of the loan.

Ampere (amp) A measure of the amount of electric current flowing through a circuit per unit of time. It is named after the French physicist André Ampère.

Anchor bolt A steel bolt with threaded head used to anchor the sill plate to the concrete or masonry foundations.

Annual interest rate The interest rate on a mortgage loan based on the nominal amount of the loan without deducting the points and financial charges.

Appraisal An estimate of the market value of a property.

Appraiser A professional trained to appraise properties.

Apron An interior trim member installed at the bottom of a window immediately below the stool.

Arbitration A procedure to settle differences or disputes between two parties through an impartial third party.

Asphalt A by-product of refining crude petroleum.

Assessed value A value placed on a property by a public officer or a board as a basis for taxation.

Assessment A charge against real property made by a branch of government to cover the proportionate cost of an improvement such as a street or sewer.

Assignee A person to whom a right or property is transferred.

Assumption of mortgage The buyer of a property assumes liability for payment of an existing mortgage loan against the property.

Attic The accessible space located between the top of the ceiling and the underside of the sloped roof.

Attorney-in-fact A person who is given

written authority by another person to sign documents on his or her behalf.

Backfill Excavated or imported material used around the foundations or in filling a trench.

Backhoe An excavating machine with a shovel at one end and a hoe at the other end.

Balloon loan A loan in which the final payment is larger than the regular monthly payments.

Baluster A vertical member that supports the stairway handrailing.

Bargain and sale deed A deed containing no specific covenants by which the owner transfers all his rights, title, and interest to the buyer.

Base molding A molding consisting of strips of wood used to cover the joints between the floor and the adjoining walls.

Baseboard heater A heat emitting device consisting of ½-inch diameter copper tube with many fins welded to it.

Batt A strip of fiberglass or rock-wool insulation precut to 4 to 8 feet long by 15 or 23 inches wide.

Batten A narrow strip of wood used to cover the joints between plywood or board panels.

Batter board A horizontal board nailed to two short posts used to establish the outside edges of the exterior walls of the house (building lines) and the elevation of the foundations.

Bay window A window projecting outward from the wall of the house.

Beam A horizontal supporting structural member made of wood or metal.

Bearing wall A wall that supports a part of the floors and roof above it.

Bench mark A mark on a permanent object indicating a verified elevation, used by surveyors as a reference point.

Binder An offer to buy real estate accompanied by a deposit of money as evidence of the buyer's good-faith intention to complete the transaction.

Blanket mortgage A mortgage loan secured by more than one piece of property.

Blankets Rolled strips of fiberglass or rock-wool insulation 15 or 23 inches in width.

Bond The evidence of a personal debt secured by a mortgage or other lien to a real property.

Boring A hole drilled in the ground to obtain undisturbed samples of the soil at various depths.

Brick veneer A facing of bricks used as exterior siding. It does not carry any load other than itself.

Bridging Small diagonal members used to brace the floor joists at their midspan.

BTU (British Thermal Units) The quantity of heat required to raise the temperature of one pound of water one degree Fahrenheit.

Building code Regulations established by state and local governments stating the minimum standards for construction.

Building line A line fixed at a certain distance from the front of a lot, beyond which no structure can project.

Building loan A short-term loan used to finance house construction. It is paid in installments during the progress of work.

Bulldozer An excavating machine on tracks (crawler), with a steel blade that can be raised or lowered attached to its front. It is used to move earth from place to place and to shape the grade.

Buy-down A technique used by developers to entice customers, whereby the developer pays a lender a sum of money to compensate it for offering buyers a lower interest rate for an early period of the life of the loan.

Cantilever A projecting beam supported at only one end.

Casing A molding used to trim door and window frames; a pipe used for well drilling.

Cement Usually refers to Portland cement. A fine gray powder that produces a bonding paste when mixed with water.

Certificate of occupancy A certificate issued by the building department stating that the house has been built in accordance with the local building code and zoning ordinance, and may be occupied.

Certificate of reduction of mortgage A document prepared and signed by the lender indicating the amount of the outstanding principal on an existing mortgage loan and the interest rate.

Certificate of title A certificate issued by a lawyer on the status of a title to real property based on his examination of the public records.

Chain of title A history of property ownership and encumbrances starting with the present owner reaching back to the original grantor, or as far back as records are available.

Circuit breaker A switch that automatically breaks the electric circuit when a short or overloading occurs.

Clay A very fine-grained inorganic soil of size less than 0.005 mm. It becomes plastic when mixed with water and loses its plasticity when it dries.

Closing The process by which the promises and agreements between the parties to a real estate transaction are fulfilled.

Closing agent The person who conducts the closing or settlement proceedings.

Closing statement (settlement statement) An account prepared by the closing agent of all the money received from and charged to both the buyer and seller.

Cloud on the title An outstanding claim or encumbrance, which if valid, would affect or impair the title to a property.

Collateral Something of value pledged as security for a debt.

Commingling The mixing of the clients' funds or deposits with the agents' personal funds.

Commission A payment due a real estate broker for services rendered.

Commitment A pledge or promise.

Competent parties Persons who are considered legally capable of entering into contract with others based on minimum age and sound mind.

Compressor The part of the air conditioning or heat pump unit that compresses the refrigerant gas so that it can absorb heat.

Concrete A construction material made of Portland cement, fine aggregate or sand, coarse aggregate or gravel, and water, mixed together.

Condemnation Taking private property for public use, with fair compensation to the owner, by exercising the power of eminent domain.

Contract for deed A contract to buy real estate by installment, whereby the buyer takes possession of the property, but the seller retains title until the price is paid in full.

Consideration Anything of value such as money or personal services that induces a person to enter into a contract. An example is what the grantor receives in exchange for his deed.

Constructive notice Notice given to the world by filing a document in the public records. The law charges the buyer with knowing what is in the public records.

Contract A legally enforceable agreement between competent parties to do or not to do certain things for a consideration.

Contractor A professional who specializes in one or more disciplines of construction such as excavation, plumbing, etc. He may be hired to provide labor only or labor and material.

Conventional mortgage loan A loan that is not FHA-insured or VA-guaranteed.

Conveyance The transfer of title to real property from one person to another, by means of a deed.

Corner bead A light-weight metal angle used to shape and reinforce outside corners in drywall, or sheetrock, construction.

Cornice Eave overhang of a pitched roof, usually consisting of soffit and facia boards.

Counteroffer An offer made in response to an offer received.

Covenant An agreement written in a deed or other document promising performance or nonperformance of certain acts, or stipulating certain uses or restrictions on use of the property.

Crawl space An enclosed shallow space below the first floor of a house without a basement.

Credit report A report prepared by a specialized company detailing an individual's debts and credit worthiness.

Damages Compensation awarded by court to a person who has sustained an injury, either to his or her person, property, or rights through the act or default of another.

Damper A valve installed inside an air duct to regulate the flow of air.

Deck A raised platform used for recreation.

Deed A written document that, when properly executed and delivered, conveys or transfers title of real property.

Deed restrictions Conditions and restrictions placed in the deed or a separately recorded declaration by a grantor for the purpose of limiting the use of the land by future owners.

Default Failure to fulfill the terms of an agreement or a contract.

Deficiency judgment A personal judgment levied against a mortgagor (borrower) if the sale price of the foreclosed upon property is less than the amount of debt.

Delay cap A device that delays explosion of the blasting charges for a few milliseconds.

Delinquent loan A loan wherein the borrower fails to make mortgage payments on time.

Delivery of deed The transfer of ownership of real estate through delivery of deed from the grantor to the grantee.

Developer-builder (developer) A person who constructs and sells houses on lots he or she owns.

Discount points An up-front fee charged by mortgage lenders to raise the effective yield on their loans. One point is equal to one percent of the value of the loan.

Dormer A vertical structure projecting from the slope of a roof and containing one or more windows.

Double-hung window A window consisting of two sashes that can slide vertically.

Downspout (leader) A vertical metallic tube, mostly aluminum, that discharges the rain and melted snow from the gutter to the ground, or to a drainage system.

Drainpipe A perforated plastic or clay pipe embedded in gravel or crushed stones and laid around the footings to drain away the subsurface water in order to keep the basement dry.

Driveway A private road from a public street to the door of the garage, or the carport.

Dry well A pile of stones buried below the elevation of the footings to collect water from the drainpipe. This water evaporates during hot weather.

Drywall (sheetrock) Sometimes called gypsum boards or plaster boards, they consist of a gypsum core faced with heavy paper on both sides, and are used to cover interior walls and ceilings.

Duct An enclosed rectangular or circular tube used to transfer hot and cold air to different parts of the house.

Earnest money A cash deposit paid by the prospective buyer of real property as evidence of his good-faith intention to complete the sale.

Easement A right or privilege that one party has in the property of another that entitles the holder to a specific limited use of the property.

Eave overhang The projection of a sloping roof beyond the exterior walls of the house.

Effective interest rate (effective yield) The interest rate based on the face value of the loan less the discount points and other financial charges.

Eminent domain The right of the federal and state governments or public service organizations to acquire all or part of a privately owned property for public use regardless of the owner's wishes. The law requires that the owner be fairly compensated.

Encroachment An unauthorized intrusion of a building or other improvement, such as a fence, upon the property of the adjoining owner, a sidewalk, or a street.

Encumbrances Any clouds against clear title of a property that diminish its value. Such clouds may include mortgage loans, liens, easements, deed restrictions, or unpaid taxes.

Engineer's chain An old U.S. unit of linear measurement. One engineer's chain = 100 feet = 100 links.

Equity The appraised market value of a property less all the debts against it.

Equity loan A junior mortgage based on a percentage of the equity in the real property and secured by a second or third mortgage.

Equity of redemption The right of the owner to reclaim his property before the foreclosure sale by paying the debt, interest, and costs of the sale.

Escrow A written agreement between two or more parties by which they deposit money or legal documents, such as earnest money or deeds, with a third party with instructions on how to disperse the documents or money.

Escrow closing A closing, or settlement, in which the buyer and seller choose a neutral third party (an escrow agent) to handle the transaction.

Estate The extent of one's rights or interest in real property.

Estoppel certificate A document signed by the mortgagor (the borrower) stating the unpaid balance of the mortgage as of the date of signing the certificate, and the interest rate on the loan.

Evidence of title Proof of ownership of a real property.

Execute To sign a document, thereby making it legally valid.

Exterior siding The covering of the exterior walls of a house. It may be made of aluminum, shingles, brick or stone veneer, or other material.

Facia The flat board enclosing the exterior ends of the rafters of the roof.

Fathom An old U.S. unit of linear measurement. One fathom = 6 feet.

Federal Deposit Insurance Corporation (FDIC) A federal agency that insures each deposit account at member banks for up to $100,000.

Federal Home Loan Mortgage Corporation (FHLMC) "Freddie Mac" is a federal agency established in 1970 to provide a secondary market for mortgage loans.

Federal Housing Administration (FHA) A federal agency that insures mortgage loans made by primary lenders against default by borrowers.

Federal National Mortgage Association (FNMA) "Fannie Mae" is the largest secondary market buyer and seller of mortgage loans made by primary lenders.

Federal Savings And Loan Insurance Corporation (FSLIC) An agency that insures each deposit account at member savings and loan associations for up to $100,000.

Fee simple (fee simple absolute or fee) The highest form of ownership of real property, continuing forever. It entitles its holder to the entire property with unconditional power of disposition during his life and descending to his distributees and legal representatives upon his death.

Fee simple determinable (qualified fee) A type of ownership that is subject to certain limitations by the grantor as indicated by the phrases "as long as," "while," or "until." If the grantee violates one of the limitations, ownership returns to the grantor automatically.

Fire blocks (fire stops) Short horizontal members nailed between the studs to prevent the spread of fire and smoke.

Fire bricks Heat resistant bricks used for lining fireplaces.

Fire underwriters A certificate issued by the town's electrical inspector indicating that the electrical wiring of the house was done in accordance with the codes and regulations.

Fireplace An enclosure to hold open fire.

It is comprised of a firebox, a throat, a damper, a smoke chamber, and an ash pit.

First mortgage A mortgage that has no recorded predecessor or lien.

Flagstones Flat pieces of grayish-blue stone of irregular size used for exterior paving, tiling, or window sills.

Flashing Rust-proof metal strips or asphalt roll used in roofing to seal the seams in the valleys between different slopes, and around the brick chimneys.

Flood insurance A federally subsidized insurance policy for flood-prone areas that must be carried by owners of properties financed by loans, grants, or guarantees issued by a federal or federally regulated agency.

Flood-plain area An area designated by the department of Housing and Urban Development (HUD) to be flood-prone.

Flue pipe A pipe that carries the smoke and exhaust gases of the fireplace or the boiler upward to the atmosphere.

Footing A concrete base poured over the bearing soil the function of which is to distribute the loads of the foundation walls above it over a wide area of soil.

Foreclosure Legal proceedings whereby a property pledged as security for a debt is sold to pay the debt in the event of the borrower's default.

Forfeiture Loss of money or anything of value as a penalty for not fulfilling a contract or an agreement.

Formica A plastic material available in different colors that is used to veneer plywood or particle wood vanities, kitchen cabinets, and counter tops. (A trade name.)

Foundation The part of the structure below the first floor or below grade the function of which is to safely support the house throughout its life. It consists of footings, walls, and slabs.

Framing Sometimes called rough framing, is a type of construction in which the studs, rafters, joists, sole plates and roof plates are put together to form the skeleton of the house.

Frost line The depth below the surface of the earth at which subsurface water may freeze during cold weather. This depth varies from region to region and is set forth by the local building codes. Footings must be poured below the frost line.

Furnace An enclosed structure in which heat is produced by burning oil, gas, or coal, or by electricity. Examples of furnaces are oil and gas boilers and electrical heating units.

Furring Thin wood or metal strips used to even the joints of walls and ceilings. Also used as a fastening base for finish material.

Gable roof A roof of triangular shape.

Gambrel roof A two-sloped roof with its lower parts steeper than its upper parts. This type of roof is identified with the Dutch Colonial house style.

Gap in title A missing link in the chain of title, such as unrecorded deed for a past period of time.

General contractor A person who contracts to build a house or building, or a part of it, for another person for a profit.

General lien A right given by court to a creditor to attach a lien to all the debtor's assets to secure the payment of a debt. A judgment lien is one example.

G-I loan (VA-guaranteed loan) A government-guaranteed mortgage loan for veterans.

Girder A large or main beam that supports numerous joists and heavy loads along its span.

Good-faith An act that is done in honesty and sincerity.

Government National Mortgage Association (GNMA) "Ginnie Mae" is a government agency and a division of HUD. It is a major secondary market mortgage trader. It also manages federally aided housing programs.

Government survey system (U.S. system of public land surveys or rectangular survey system) A method of describing land in the 30 states that include public land, whereby the land is divided into squares of various sizes.

Grace period Additional time allowed to make a payment without being considered in default.

Grade The ground level or elevation. Also the slope of the surface of a lot or a road.

Graduated payment mortgage A mortgage in which the initial monthly payment is less than the interest. The payment increases by a certain percentage each year for a specific number of years, then remains at that level for the remaining term of the loan.

Grain The direction of fibers in a piece of wood.

Grant A term used in deeds to indicate the act of transferring title to real property.

Grantee The person to whom the title to real property is transferred (the buyer).

Grantor The person who transfers title to real property by deed (the seller).

Grantor-grantee indexes Two sets of public record books in which all recorded real property documents are listed in alphabetical order—one according to the names of the grantors and mortgagors and the other according to the names of the grantees and mortgagees.

Grout A mortar made of sand and cement or other ready-made mix used to fill the joints between tiles; between the old and new surface of poured concrete; and around anchor bolts.

Gunter's chain An old U.S. unit of linear measurement. One Gunter's chain = 66 feet.

Gutters Mostly aluminum, troughs installed along the facia to collect the rain and melted snow from the roof.

Gypsum board See **Drywall**.

Habendum clause A clause in the deed beginning with the words ''to have and to hold,'' stating the extent of ownership the grantor is transferring.

Hardboard Boards made by compressing shredded wood chips or fibers with a bonding substance.

Hazard insurance An insurance policy for occupied houses that insures against losses resulting from hazards such as fire, windstorms, and flooding. It does not cover flooding in flood-prone areas so designated by HUD.

Header Heavy or double framing laid at the top of windows, doors, or other big openings; the vertical enclosure installed on the outer edge of the sill plate.

Hearth The floor of a fireplace. The **front hearth** is the elevated area in front of the fireplace.

Heat gain Heat transferred into the house from the outside during hot weather through windows, walls, exterior doors, and ceilings.

Heat loss Heat escaping from the heated house to the outside during cold weather through windows, walls, exterior doors, and ceilings.

Heat pump An electric unit that cools the house during hot weather by absorbing heat from inside and discharging it to the outside. In cold weather, it absorbs heat from outside and discharges it inside.

Heirs and assigns Terms used in deeds. Heirs inherit the property when the owner dies; assigns are the persons to whom the owner transfers the title of the property.

Hip roof A roof with sloping ends and sides.

HUD The U.S. Department of Housing and Urban Development. It prepared a booklet titled ''A HUD Guide for Home Buyers—Settlement Costs,'' a copy of which must be given by lenders to their loan applicants.

Humidifier A device for raising the indoor humidity to a comfortable level. Hot air heating necessitates humidifying.

Improvements Additions made to the property. This includes the house, grading, pavements, sidewalks, sewer and water lines, and fences.

Initial point An established and secured

mark used as a reference in the government survey system.

Inspection Examining the components of the house such as the framing, electrical wiring, plumbing, walls, attic, basement, boilers, etc. to discover any defects.

Installment contract See **Contract for deed.**

Institutional Lenders Financial institutions such as savings and loan associations, commercial banks, mutual saving banks, mortgage companies, etc. whose lending practices are regulated by federal or state laws.

Instrument A formal legal document such as a deed, mortgage, promissory note, or contract.

Insulation Material installed in the walls and ceiling to conserve energy. It may be made of fiberglass, rock-wool, cellulose, or vermiculite.

Interest The amount that lenders charge borrowers for using their money. The most widely used method of charging interest is the annual interest rate, which is the yearly charge stated in the form of a percentage of the loan.

Jambs The pieces forming the sides and top of a window or a door frame.

Joint compound A compound used in drywall or sheetrock installation to gloss over the paper tape and nail heads to conceal the joint and produce a smooth surface.

Joists Horizontal beams that support a floor or ceiling.

Judgment lien A lien on the properties of a debtor to secure the payment of a court award for a claim, or as damages.

Junior mortgage A mortgage which has a prior recorded mortgage or lien.

Land The surface of the earth extending downward to the center of the earth and upward to infinity.

Land contract See **Contract for deed.**

Land description A description that positively identifies a piece of property. Courts consider a land description to be legal if it enables a land surveyor to locate the property.

Landing A platform at the beginning and end of a flight of stairs.

Landscaping The shrubs, bushes, and trees planted around the house in an organized manner designed to enhance the property.

Late charge A penalty that a lender imposes on a borrower if he fails to pay a loan installment on time.

Lath Wood, metal, or gypsum boards that are fastened to the wood frame to form a base for plastering or tiling.

Lawyer's opinion of title A written report on the condition of title prepared by a lawyer after he examines an abstract of title.

Leveling rod A graduated wood or metal rod of rectangular cross section used to measure the difference in elevation between two points.

Lien A legal right or claim that one party (lienor, mortgagee) attaches to the property of another (lienee, mortgagor) as security for paying a debt or obligation.

Lien-theory states States that consider a mortgage to be merely a lien to the property.

Life estate Ownership of a property that is limited in duration to the life of the grantor, grantee, or a third party.

Line of credit The maximum amount a bank will lend through a revolving credit account whereby the borrower is free to borrow and repay any amount provided that the outstanding debt does not at any time exceed the line of credit. Some commercial banks offer large lines of credit secured by mortgages.

Lis pendens A notice filed in the public records where the property is located informing the public that a legal action against a property is pending.

Listing An agreement between the seller and a real estate broker by which the broker is authorized to represent the seller in soliciting offers to buy the property. This is in return for a negotiated fee.

Live load Any load that is not the actual weight of the structure itself. It includes people, furniture, books, wind and snow.

Loan-to-value ratio The ratio between the amount of the principal of a mortgage loan and the appraised market value of the property or the purchase price, whichever is lower.

Lot Also known as a parcel or property, is a piece of land that may be used as a building site for a house.

Love and affection In itself sufficient consideration (price) for a real property when it is conveyed by deed to a relative as a gift.

Lumber boards Wood that is less than two inches thick and two or more inches wide.

Lump sum Sometimes called fixed price; a method of payment for construction work whereby the contractor agrees to do specific work for a specific amount of money.

Mansard roof A roof style of Italian origin, popularized in France.

Mantel The decorative framing around a fireplace.

Maps and plats Land survey of a subdivision including several lots showing the boundary lines and measurements of each lot, and the streets, utilities, and the monuments of the subdivision. It must be prepared by a licensed land surveyor or a civil engineer.

Marketable title A title that does not involve the buyer in litigation. Also, a title that a court could compel the buyer to accept.

Market value The highest price a property is likely to command if allowed to stay on the market for a reasonable amount of time.

Mason A professional who builds brick, stone, or concrete work.

Masonry Structures constructed of bricks, stones, or concrete.

Material supplier (materialman) A supplier of materials to be used for construction (i.e., sand, gravel, and bricks).

Maturity The date by which a mortgage loan must be paid in full.

Mechanic's lien A right given to laborers, material suppliers, contractors, and their subcontractors to secure payment for either work performed or material furnished where the value or condition of the property has been improved and the workers and material suppliers have not been paid.

Meeting of the minds Mutual agreement between the buyer and the seller regarding the provisions of a contract.

Meridian The direction of an imaginary line running north–south used by surveyors as a reference in defining the directions of the boundary lines of a property. The most commonly used meridian is the true, or astronomic, which is the direction of a line passing through the earth's geographic poles.

Meter The basic SI (International System) unit of linear measurement. One meter = 39.37 inches.

Metes-and-bounds An old method of describing land by specifying the measures, directions and shapes of the property lines with reference to roads, rivers, and monuments.

Millisecond One thousandth of a second.

Miter Carpenter's terminology for a joint formed by leveling the edges of two pieces of wood at 45 degree angles and fitting them together.

Monuments A natural or man-made fixed mark or object that can be easily identified. They are used to define property lines in the metes-and-bounds method of land description.

Mortar A mixture that may be made of cement, lime, sand, and water to form a plastic bonding agent for brick laying or tile setting. It hardens after a few hours holding the bricks or tiles together.

Mortgage A legal document by which the borrower pledges his ownership of the property as collateral to secure the mortgage debt included in the promissory note, note, or bond.

Mortgage banking companies Sometimes called mortgage companies, they provide mortgage loans by either acting as liaison for big investors or by financing the loans from their own funds, then selling them quickly in the secondary market.

Mortgagee The lender. The one who receives and holds a mortgage as security for the payment of the debt.

Mortgage Guarantee Insurance Corporation (MGIC) A major privately owned corporation that insures the top 20 or 25 percent of mortgage loans made with 10 or 5 percent down payments, respectively.

Mortgage insurance An insurance that pays the monthly installments of a mortgage loan if the borrower fails to make the payments.

Mortgage life insurance An insurance that pays the balance of the mortgage loan if the borrower or co-borrower dies.

Mortgagor The borrower. The one who gives a mortgage as security for the payment of the debt.

Mosaic tile A tile with a top layer made of crushed pieces of colored marble or stones mixed with white or colored cement.

Mullion A vertical piece of framing between two adjoining windows.

Multiple unit dwelling A house that includes several separate living units, each accommodating one family. (i.e., a two-family, or four-family dwelling.)

Multiple listing service (MLS) A real estate brokerage service organized by a group of brokers within a geographical area to make any member's listing available to the other members.

Mutual savings banks Saving institutions mutually owned by their depositors. They are state chartered.

Negative amortization The balance of the principal of the mortgage loan increases each time a payment is made. This oc-

curs when the monthly payment is less than the interest.

Newel post A main post that supports the stair railing at either end.

Nosing The extension of a stair tread beyond the face of the riser below.

Notary public A public officer who certifies signatures, attests to or certifies written documents, administers oaths, and takes acknowledgments of deeds and other mortgage documents.

Note (promissory note) A document signed by the borrower that includes the amount of the loan, the interest rate, the method of repayment, and the borrower's promise to repay the loan plus interest to the lender. The note must include a clause stating that it is secured by a mortgage.

O.C. (on-center) The distance between the centers of two consecutive members in equally spaced framings, or reinforcing bars.

Offer A promise by a party to do a specific act or give something to another party in return for a specific act or a promise.

Offer and acceptance See **Meeting of the minds.**

Offer to purchase See **Binder.**

Open-end mortgage A mortgage in which the borrower can obtain additional advances from an existing amortized loan for up to the amount that has been amortized or a top limit stated in the loan agreement.

Origination fee A fee charged by lenders to cover their administrative costs in processing mortgage loans, often expressed as a percentage of the loan.

Package mortgage A mortgage that includes all appliances such as the refrigerator, washer, dryer, etc. The borrower cannot sell any of these items without the lender's consent.

Panel A flat rectangular piece of building material such as hardwood, plywood, or drywall. Also, a rectangular piece of glass or wood set in a frame.

Parcel A lot; a piece of land.

Parquet floor A floor made of short pieces of hardwood laid in different design patterns.

Partial release clause A clause usually included in a blanket mortgage by which a lot can be released by repaying a specific portion of the debt.

Partially amortized loan A loan that is repaid by a specific number of amortized payments followed by a balloon payment at the end of the loan's term.

Partition An interior dividing wall.

Party wall A wall constructed on or at the party line between two adjoining lots to serve as the exterior wall for the adjoining houses.

Patio A recreational area constructed on the ground.

Performance bond A surety bond posted by a contractor for the owner's benefit to assure him that the work will be completed in accordance with the contract agreement. It is issued by a surety company in an amount in excess of the value stated in the contract.

Pier A vertical support for a structure formed by drilling a hole in the earth to below the frost line or as specified in the design drawing then filling it with concrete or bricks.

Pitch The slope of a roof or a pipe.

PITI Abbreviation for principal, interest, taxes, and insurance, which constitute the monthly housing costs. Most lenders require that the value of PITI should not exceed a percentage of the borrower's gross monthly income.

Plans (blueprints) The design drawings of a house showing all the elevations, plans and details needed to construct the house.

Plastering Applying one or more coats of plaster on walls and ceilings to produce a smooth surface.

Plasterboard See **Drywall**.

Plat of subdivision A map of a subdivision indicating the block numbers; the lo-cation, boundary lines, dimensions and number of each lot; and the location and names of the existing and planned streets.

Plumb Perfectly vertical.

Plywood A panel consisting of thin layers of wood glued together with the grain direction of each layer perpendicular to those of the adjoining layers.

Point of beginning (POB) The starting point in the metes-and-bounds method of land description. It must be a permanent well-referenced corner of the property.

Polyurethane finish A clear finish used for coating stained wood to provide it with protection and shine. It is durable and highly resistant to water.

Porch A covered recreational area with a roof separate from that of the house; a veranda.

Portico A covered entrance of a house, usually supported by decorative columns.

Post A vertical supporting member.

Power of sale The right given by lenders in some states to sell the foreclosed upon property without obtaining a court order.

Prepayment penalty A charge set by lenders for repaying part of or all of the debt before its maturity.

Pressure-treated wood Lumber that has been saturated with a preservative under pressure causing it to penetrate the pores. This wood is used for outdoor structures such as decks and porches.

Primary mortgage market Includes all the institutions that lend directly to mortgage borrowers.

Primer-sealer The first coat in painting a new surface. It seals the pores of the raw surface and bonds well to the finish coats.

Principal The amount of the mortgage debt; a main party in a transaction; the homeowner who hires a real estate broker.

Principal meridian A true north–south line that passes through the initial point and is used as a reference in the government survey system.

Priority of liens The order in which the liens are to be paid if the property is foreclosed upon. The priority is generally in the order in which the liens are recorded.

Private mortgage insurance (PMI) A policy provided by private companies that insures the top 20 or 25 percent of small down payment mortgage loans made with loan-to-equity ratios of 90 and 95 percent, respectively.

Promissory note See **Note.**

Property A piece of land; a lot; the rights and interests a person has in a thing he or she owns.

Purchase-money mortgage A mortgage that the buyer gives the seller for a part of the price of the property.

Quadrangle (tract) The largest unit of area in the government survey system measuring 24 miles square.

Qualified fee See **Fee simple determinable.**

Quarter section A unit of land measurement in the government survey system equal to one quarter of a square mile or 2640 feet square or 160 acres.

Quasi Similar to; to some degree.

Quiet enjoyment The right of an owner who is legally in possession of the property to use it without interruption from any person claiming to have a superior title to the property.

Quitclaim deed A deed the usual purpose of which is to release any interest a claimant may have in the property. It includes the words "remise, release, and quitclaim," instead of the words "grant and release," used in other types of deeds.

R-value A measure of the resistance of a building material to heat flow. Higher R-values indicate better thermal insulating characteristics.

Rafters Sloping beams supporting a roof.

They extend from the central ridge board to the edge of the eave overhang.

Rail The horizontal top, bottom, and middle framing pieces of a panel door. A top or middle bar extending over or between posts.

Real estate The land and all the improvements on it including the house; synonymous with real property.

Real Estate Settlement Procedures Act (RESPA) A federal law passed by Congress in the mid-1970s requiring lenders to provide their loan applicants with good-faith estimates of settlement or closing costs. RESPA regulations apply to first mortgage loans only.

Real property See **Real estate.**

Recording The act of filing documents affecting real property such as deed, mortgages, and liens in the public records of the county where the property is located. Recorded documents must be signed and acknowledged.

Rectangular survey system See **Government survey system.**

Redemption period The period of time after the date of a foreclosure sale during which the property owner has the right to redeem or reclaim his property by paying the debt, interest, and sale costs. This period varies from state to state.

Reduction certificate See **Certificate of reduction of mortgage.**

Refinance Obtaining a new mortgage loan to repay an existing one.

Referee's deed Conveys property sold by court order in a foreclosure proceeding.

Registrar The public officer in charge of maintaining the public records (in some parts of the country).

Regulation Z (Truth-in-Lending Act) A federal law that requires lenders to provide prospective borrowers with a statement disclosing the annual interest rate, the effective interest rate or effective yield, and an estimate of the prepaid charges.

Release clause See **Partial release clause.**

Rescind Cancel.

Rider An addition or amendment to a document.

Ridge board The highest beam, which supports the upper end of the rafters of a roof.

Right to rescission The right of a borrower to cancel a loan transaction that is secured by a mortgage or lien on a home he occupies as his personal residence.

Rise The vertical distance between two points.

Riser The vertical piece between two successive stair treads.

Run with the land Rights that are binding to successive owners of the property such as easements granted to a utility, and deed restrictions.

Saddle A strip made of wood, marble, or metal installed at the step of a door or arched opening to form a boundary for flooring materials such as tiles, wood, and carpeting.

Salt box roof A side gable with one side long and extending down to the first floor and the other side short covering the second floor.

Sash The frame in which a pane or panes of glass are set in a window; the movable part of a window.

Satisfaction of mortgage A certificate issued by the lender stating that the mortgage has been paid in full. It describes the mortgage fully and indicates where it is recorded. The borrower must record this certificate as soon as possible to clear the title.

Savings and loan associations The largest home mortgage lenders.

Second mortgage A mortgage to a property that already has a first mortgage. It is also called a junior or subordinate mortgage.

Secondary mortgage market Includes large institutions that provide cash to the housing market by buying mortgage loans from lenders (primary market), and packaging and selling them to large investors.

Section A unit of land measurement in the government survey system equal to one square mile or 640 acres.

Septic tank A settling tank in which the sludge in the household sewage settles and the effluent discharges into an absorption field or seepage pit.

Setback The minimum distance between the street and the building line. It is established by the local zoning ordinances or deed restrictions.

Settlement The downward movement of the soil supporting the footings; also see **Closing.**

Shakes Shingles made of hand split wood used for roofing and exterior siding.

Sheathing Boards or plywood or particle wood panels used to cover the wood framing of the roof and exterior walls.

Shed roof A roof sloped in one direction only.

Shelf A recess in the top of the foundation wall that forms a base for the brick or stone veneer wall.

Sheriff's deed Deed given by court to the buyer of a property through a foreclosure sale administered by a sheriff.

Shim Thin strip of metal, wood, or stone used to fill a gap or level a surface.

Shingles Small thin pieces of building materials such as asphalt, wood, or slate used to cover the roof and exterior walls.

SI (Système International d'unités) The international system of unit measurements intended for use throughout the world and of which the U.S. is a member. The basic unit of linear measurement is the meter, which is equal to 39.37 inches; the unit of area measurement is the hectare, which is equal to 2.471 acres.

Sideyard The minimum distance between the sides of a house and the property lines facing them, as set forth in zoning ordinances or deed restrictions.

Sill The horizontal member forming the base of a window, or the foot of a door.

Sill plate Pieces of lumber laid on top of the foundation walls to form a base for the wood frame.

Single-family house A house that is designed and certified for occupancy by one family only.

Stain A die used for finishing wood surfaces.

Skylight An opening in the roof covered with thick glass the function of which is to light the area below it.

Slab A thick plate made of concrete.

Soffit The boards covering the underside of the eave overhang.

Soil The loose upper layer of the earth. The bearing soil is the layer of soil on which the footings are poured.

Solid blocking Solid members placed between adjacent floor joists to prevent them from twisting and to distribute concentrated loads imposed on any joist.

Span The distance between the structural supports of beams and girders.

Specifications Detailed, precise engineering instructions that include the kinds of materials to be used and the method of construction.

Specific performance Court action to compel a party to fulfill the terms of a contract.

Square 100 square feet of roofing material.

Stakes Pieces of wood inserted in the ground at the corners and along the boundary lines of a piece of property to precisely define its boundaries.

Staking Marking the corners and boundary lines of a property and the corners and building lines of a house by means of stakes.

Statute of frauds A state law requiring that certain real estate documents, such as contract of sale, mortgages, and easements, must be in writing in order to be enforceable.

Statute of limitations A law setting the maximum period of time during which a lawsuit may be filed after the occurrence of the cause of the suit.

Statutory redemption A right given by

most states to a borrower to redeem his property after a foreclosure sale.

Stile The vertical side pieces of a panel door.

Stool A flat molding installed at the bottom of the window over the apron.

Stringers The sloping sides of a stair flight that support the treads and risers; small beams.

Stucco A covering for exterior and interior walls applied in the form of plastic mortar or paste that hardens in a few days. The paste mix for exterior walls consists of Portland cement, sand and lime. Interior walls may be covered by the same mix, or by fine plaster.

Studs The vertical members in wall framings. Their function is to form the walls and support the floors and the roof above them.

Subdivision Land that is divided or designated to be divided into two or more lots.

Subflooring Plywood or boards nailed directly to the floor joists to form a base for the finish flooring.

Subordination clause A clause by which a second mortgage lender agrees that his mortgage will not become a first mortgage but will remain a second mortgage if the first mortgage is refinanced.

Subsurface water Water below ground that is caused by heavy rainfall.

Sump A square, rectangular, or circular pit used to collect water, usually from a wet basement. The collected water is subsequently discharged by a pump.

Surety A party that assumes liability if another party fails to perform an agreement, fulfill a contract, or pay a debt.

Surety bond Insurance guaranteeing performance of a contract or obligation.

Survey A drawing made to scale showing the lengths and directions of the boundary lines of the lot; the surrounding lots and streets; the position of the house and all exterior improvements such as walkways, driveways, decks and porticos within the lot; and any existing encroachments.

Tandem plan A mortgage subsidy program administered by the Government National Mortgage Association whereby it buys mortgage loans made to low income families at below-market interest rates with the U.S. government absorbing the difference between these rates and the prevailing rates.

Taping The process of covering the drywall joints with paper tapes and glossing over them with several coats of joint compound to produce a continuous smooth surface.

Tax deed A deed conveying title to a property that has been sold by a government authority due to a failure to pay taxes.

Term The length of time by the end of which the mortgage loan must be paid in full (i.e., a 30-year term mortgage).

Theodolite An instrument used by surveyors. It consists of a telescope that can rotate vertically or horizontally to measure horizontal and vertical angles with great accuracy.

Time is of the essence A clause used if either the buyer or the seller of real property has compelling reasons for closing the title on a fixed date stated in the contract.

Title The right of ownership to real property.

Title insurance policy A policy issued by a title insurance company after it searches the public record. It insures against losses resulting from undiscovered defects such as a forged document, incompetent grantor, incorrect marital status, or improperly recorded deed.

Title search The examination of the public records to discover the names of the parties who have interest in a real property and to detect any defects that may affect the quality of the title.

Title theory states States in which the law considers that a mortgage transfers some title rights from borrower to lender.

Top plates Pieces of lumber laid horizontally on top of the studs to tie them together and to form a base for the framing above which may be a floor or a roof.

Topography A drawing that indicates the configuration of the earth's surface and the locations of the natural or man-made monuments.

Torrens certificate of title registration A valid proof of ownership in several states revealing the name of the owner, type of ownership, and all the liens and encumbrances attached to the title. Transfer of ownership does not require title search.

Township A unit of area measurement in the government survey system equal to six miles square.

Tract A piece of property; a lot; a large area that can be subdivided into several lots.

Transit See **Theodolite**.

Tread The horizontal surface of a stair that people step on when climbing up or down.

Trim Finish carpentry, including installation of interior doors, window and door casings, closets, shelves, molding and hardware.

Truth-in-lending act See **Regulation Z**.

Underlayment A layer of material placed under carpeting, resilient flooring, or roof shingles.

Uniform settlement statement See **Closing statement**.

U.S. system of public land surveys See **Government survey system**.

Usury Interest charged in excess of the maximum legal interest rate allowed by the law of the state.

Valley The intersection of two sloping roofs.

Valuable consideration The price paid for the purchase of a real property. It may be in the form of money or services.

Vapor barrier A waterproof membrane or a plastic sheet laid below slabs poured on the earth and on the side of insulation facing the heated area to control dampness and condensation.

Vara An old Spanish unit of linear measurement equal to about 33 inches, often encountered in the southern and southwestern states of the U.S.

Veneer A protective or decorative nonstructural facing, such as brick, stone or formica veneer.

Warranty deed (warranty deed with full covenants) Warrants that the grantor (seller) has a good title free and clear of all liens and encumbrances; defends the grantee (buyer) against all claims.

Water table The top level of the natural underground water resulting from a nearby stream, drain, or shallow rock formation.

Wind load The pressure exerted on a building by wind, expressed in pounds per square foot.

Wood preservative A clear or semitransparent coating used on wood to show the grain.

Zoning ordinances Local regulations affecting property uses and type of construction. They set forth the minimum lot sizes; number of family units in each dwelling; the maximum height of buildings; and the minimum setbacks and side yards.

Index